ENVIRONMENTAL REMEDIATION TECHNOLOGIES, REGULATIONS AND SAFETY

DRIP IRRIGATION

TECHNOLOGY, MANAGEMENT AND EFFICIENCY

ENVIRONMENTAL REMEDIATION TECHNOLOGIES, REGULATIONS AND SAFETY

Additional books in this series can be found on Nova's website under the Series tab.

Additional e-books in this series can be found on Nova's website under the e-book tab.

ENVIRONMENTAL REMEDIATION TECHNOLOGIES, REGULATIONS AND SAFETY

DRIP IRRIGATION

TECHNOLOGY, MANAGEMENT AND EFFICIENCY

ALFRED H. STEELE
EDITOR

nova publishers
New York

Library of Congress Cataloging-in-Publication Data

Drip irrigation : technology, management, and efficiency / editor, Alfred H. Steele.
pages cm. -- (Environmental remediation technologies, regulations and safety)
Includes index.
ISBN 978-1-63463-737-4 (hardcover)
1. Microirrigation. I. Steele, Alfred H.
S619.T74D776 2015
631.5'87--dc23
2014047189

Published by Nova Science Publishers, Inc. † New York

CONTENTS

PREFACE

Although over two thirds of the Earth's surface is covered by water, more than 97% is ocean water which is too salty for human use or even for irrigation. Consequently, the freshwater is only 3% and almost 1% of the available freshwater is liquid surface water that can be used directly by humans. The rest is groundwater and iced water. Yet still, freshwater is considered to be one of the most abundant resources on earth. In the agriculture sector only, more than two-thirds of the available freshwater is used for irrigation. This book focuses on the technology, management and efficiency of drip irrigation.

Chapter 1 – Citrus water use is governed largely by canopy size, leaf area, root architecture and distribution, water management practices and prevailing environmental conditions such as soil characteristics and climate. Prior to 1970, citrus has been cultivated using surface and microsprinkler irrigation systems. Microsprinkler irrigation still dominates citrus production in some areas, but in the past 40 years, drip irrigation has emerged as one top technology for promoting water and nutrient use efficiency in many cropping systems including citrus. Drip irrigated area for all crops expanded 50-fold from the 1970 to 2001. Yield gains and water savings of drip over conventional surface irrigation methods have been reported to be in the order of 11 to 97% and 13 to 514%, respectively, for several horticultural crops. Advantages of drip irrigation include the ability to apply small amounts of water directly to the root zone, better nutrient management through fertigation, effective weed and pest control, and lower long-term investment and operating costs. This paper elucidates the global impact of drip irrigation and environmental conditions on citrus water use. Examples of the associated tree and soil characteristics that affect citrus water use are discussed to provide a framework for improving water management in citrus production.

Chapter 2 – The traditional scarcity of hydric resources in Spain linked with a growing trend in water demand for applications such as landscaping, gardening or agricultural irrigation is gradually shifting water management practices towards the use of reclaimed water obtained from wastewater treatment plants. From a sustainable point of view, reclaimed water irrigation must be not only sanitary safe, but also should avoid problems in irrigation networks. Drip irrigation allow us to prevent sanitary risk, even when using cheaper and less intensive water treatments. But, on the other hand, this irrigation system is more sensitive to obstruction problems. Most of the problems are related to biofilm formation at the inner surfaces as well as to the points of delivery, and a crucial step for any antifouling strategy is to characterize the microbial communities involved in its development. In this chapter we present the molecular characterization of the microbial assemblages in biofilms from drip irrigation systems in Gran Canaria (Spain) using treated wastewater from two wastewater treatment plants (WWTP), Bocabarranco and Cardones, presenting low and medium content of suspended solids (22.8 and 30.6 $mg \cdot L^{-1}$) respectively. Different samples (new drippers and older ones with more than 5 years of use) for irrigation of road green zones were examined from clogged and unclogged drippers. They were analyzed by denaturing gradient gel electrophoresis (DGGE) and subsequent sequencing of amplified 16S rRNA gene bands. DGGE patterns showed the presence of a large number of bands, indicating that the microbial community was rather complex. Most of the retrieved DGGE bands belonged to the phylogenetic groups Alpha and Gamma-proteobacteria, Bacteroidetes, Acidobacteria, Chlorobi, Nitrospirae and Firmicutes, but sequences from Beta and Delta-proteobacteria, Actinobacteria, Chloroflexi, Planctomycetes, Gemmatimonadetes, Deinococcus-Thermus and Cyanobacteria were also found. Our results demonstrated that biofilms diversity was related with clogging phenomena, as higher diversity was found in the case of higher obstruction problems. Besides, there was an influence of dripper design on emitter obstruction (integral drippers were less prone to be clogged), a fact favoured by the presence of certain genera of bacteria. Besides, the results showed that the clogging process was able to occur in a short period of time in sensitive drippers. In general, the biofilms formed in these irrigation systems using treated wastewater were more complex than we initially expected, with representation of groups usually found in aquatic environments. They showed a high diversity of unknown and uncultured microorganisms, suggesting that well structured biofilms resistant to pressure flushing could develop.

Chapter 3 – With no doubt, polymers applications have a great influence on the way of our life because of their ease of processing that meet the specific needs. In agriculture, polymers are widely used to protect the environment using new techniques of applications. In particular, the polymeric materials are used into agricultural soils to improve aeration, conserve moisture, provide mulch, and to promote the fertility and the health of the soil. Consequently, polymers are receiving a great deal of attention as soil conditioners. Here, we focus on soil-polymer applications under surface- and subsurface-drip irrigation systems to manage the irrigation water and to improve the soil properties such as stability and water holding capacity. In this chapter, we highlight the main advantages and the drawbacks of surface- and subsurface-drip irrigation systems that use polymer conditioners in view of system construction, quality of irrigation water, evaporation and percolation, crop production, environmental regulation and safety. We also discuss in details the effects of polymer type (synthetic and natural), polymer concentration and polymer formulation process on soil properties, in view of water-holding capacity, water use efficiency, soil permeability and infiltration rates, and plant performance especially in structureless soils. In addition, we illustrate the effects of soils and water properties on polymer absorbency of water. Moreover, we show the synthesis and the characterization of the different absorbent polymeric materials and how parameters such as initiator type, initiator concentration, and polymerization type affect the polymer capacity of water absorptivity. Subsequently, we throw some light on the future impacts of the polymeric materials in the agricultuel area.

Chapter 4 – Drip irrigation is a system in which water is supplied with a given flow rate directly to plant roots to meet its water requirement. Due to water scarcity worldwide drip irrigation practice is gaining momentum because of its better control on water volume and management strategy with time and space. The system is compatible for a wide range of crop variety, soil type, climate and landscape despite of few inherent constrains. Since long-back clogging and emission non-uniformity have been the major obstacles in the development of drip irrigation using various emitters. Clogging would be a serious problem in areas with brackish water where the problems of precipitation of calcium carbonate, organic materials and suspended sands are severe. Installation of filter equipment to the system could not eliminate the problem entirely and thus irrigators have to use different acidic solutions to remove precipitation, which have adverse influence on soil and crops. On the other hand to obtain the best emission uniformity (EU) in plain/uneven lands the pressure regulators and pressure compensating emitters have been in use for a long time. However, pressure compensating emitters tend to be more complex and expensive than non-compensating emitters. This chapter suggests the possibility of utilizing small diameter pipes approximately 2 to 4mm called microtubes, for enhancing the emission uniformity of the drip system and reducing the difficulties encountered by those emitters due to clogging and blockage. Microtubes have many advantages compared to traditional types of emitters in terms of cost and practical applications. As these pipes are made of flexible materials can be adjusted in shape and length easily. By adjusting the microtube lengths according to energy head developed throughout a network, an equal emission can be delivered to any landscape. Here microtubes act as emitters along the lengths of laterals, where laterals are emerged from manifold to form a subunit of the irrigation system. The variation of the microtube lengths along laterals are done for compensating the energy heads above a threshold value in a given manifold (or subunit). This threshold value has been set at the very last microtube of the end-lateral, which is equal to the frictional and other minor head loses calculated for a minimum length microtube to reach the plant root. The lengths of the other microtubes would consistently increase according to excess heads above the threshold head would have generated. Hence these laterals can also be imagined as larger emitters along the manifold of an irrigation subunit. These larger emitters along the manifold would have a characteristic pressure-discharge relationship for a hydraulically calculated set of microtube lengths emitting equal discharges along the end-lateral. At this stage there would have two options for the design of the succeeding laterals. In one option, the length of all the microtubes in the succeeding laterals will have the same set of lengths as has been calculated for the end-lateral. This configuration of the microtube setup will deliver to some extent a variable discharge; nonetheless the designer can adopt an EU value (say EU $\geq$ 90%) to allow the maximum variation by restricting the total number of laterals and thereby the length of each manifold. In other option, the designer will vary all the microtube lengths in all the successive laterals so as to deliver equal flows and thereby to achieve full emission uniformity (EU). When the required discharge and diameters of microtube, lateral and manifold and some ground conditions are given, the length of the microtubes, the pressure heads, EU (in percentage) and the best subunit dimensions can be obtained by using this simple algorithm. The results of this study showed that larger sized microtubes would have higher variation of length, while applying larger flow rates could have decreased the microtube length variation along the laterals.

Chapter 5 – This work was carried out in northern Tunisia (36.5ºN, 10.2ºE) in order to study the behavior of local and foreign olive varieties (Olea europaea L.) cultivated in the same area under drip irrigation. Through this experiment we entend to evaluate the ability of these cultivars (collection) to valorize water. For this purpose, the water use efficiency was determined for both fruit growth ($WUE_{fr,}$ mm of fruit increase / m^3 I+P) and olive production (kg of fruits / m^3 I+P). During the three years of monitoring, 2010-2012 and 2013, trees were irrigated from May to September twice a week. Each tree received an amount of water equal to 5.7 m^3 in 2010 and 3.4 m^3 in 2012 and 2013.

Results showed that fruits grew continuously from fruit set to harvest with different trends following the season and variety even during pit hardening. Four models of fruit growth were identified, with one or two peak values depending on cultivars. Most growth (47%) occurred during the early stage of fruit development (Stage 1).

This resulted in high WUE_{fr}, of 12.1 mm/m^3 in 2010, 9.6 mm/m^3 in 2012 and 6.9 mm/m^3 in 2013. During stage 2, fruit growth slowed and some diameter shrinkage was recorded for cvs., Koroneiki (-1.4%), Branquita (-7.7%), Sayali (-20.7%) and Lucques (-22.1%) due to water shortage. Some varieties showed high rates of growth during stage 3. For the overall growing cycle, the average water use efficiencies reached 1.0 mm/m^3 in 2010, 4.2 mm/m^3 in 2012 and 3.3 mm/m^3 in 2013, with important differences between cultivars (0.20-6.22 mm/m^3). Maximums of up to 2 mm/m^3 were obtained for cvs., Barouni and Madurel, and the minimums were observed for cvs., Koroneiki, Arbequina, Chemlali and Coratina. Olive production varied consistenly between varieties and trees of the same cv.,. Average WUE_p were equal to 1.49 kg/m^3 in 2010, 2.02 kg/m^3 in 2012 and 3.28 kg/m^3 in 2013, with maximums observed for cvs., Chemlali (2.11 kg/m^3 in 2010 and 10.06 kg/m^3 in 2012) and Calegua (7.74 kg/m^3). The lowest values of water use efficiencies for olive production were recorded for cvs., Arbequina in 2010 (0.72 kg/m^3), Manzanille in 2012 (0.03 kg/m^3) and Sayali in 2013 (0.59 kg/m^3).

To conclude, we can say that most varieties valorize water more efficiently during their early fruit growth process, i.e., just after fruit set, with however some exceptions and large disparities between cultivars. This should be taken into account when planting multi-varietal olive orchards in order to get harmonious multi-varietal plantations. In such 'orchards', particular attention should be given to tree vigor and soil coverage, which was found to be correlated to WUE and olive production.

Values of WUE for fruit growth and olive production presented herein can be used indicatively for this purpose depending on the objective of planting such olive groves: WUE_{fr} is usefull for table olive orchards, while WUE_p is more suitable for oil production ones.

In: Drip Irrigation
Editor: Alfred H. Steele
ISBN: 978-1-63463-737-4

Chapter 1

Environmental Conditions and Irrigation Practices Affect Citrus Water Use

Kelly T. Morgan and Davie M. Kadyampakeni**
University of Florida, Southwest Florida Research and Education Center, Immokalee, US

Abstract

Citrus water use is governed largely by canopy size, leaf area, root architecture and distribution, water management practices and prevailing environmental conditions such as soil characteristics and climate. Prior to 1970, citrus has been cultivated using surface and microsprinkler irrigation systems. Microsprinkler irrigation still dominates citrus production in some areas, but in the past 40 years, drip irrigation has emerged as one top technology for promoting water and nutrient use efficiency in many cropping systems including citrus. Drip irrigated area for all crops expanded 50-fold from the 1970 to 2001. Yield gains and water savings of drip over conventional surface irrigation methods have been reported to be in the order of 11 to 97% and 13 to 514%, respectively, for several horticultural crops. Advantages of drip irrigation include the ability to apply small amounts of water directly to the root zone, better nutrient management through fertigation, effective weed and pest control, and lower long-term investment and operating costs. This paper elucidates the global impact of drip irrigation and environmental conditions on citrus water use. Examples of the associated tree and soil characteristics that affect citrus water use are discussed to provide a framework for improving water management in citrus production.

Keywords: *Citrus sinensis*, drip fertigation, micro-irrigation

* Current address: International Water Management Institute, PMB CT 112, Cantonments, Accra, Ghana. Email: dakadyampakeni@yahoo.com

INTRODUCTION

Plant Water Requirements

In general, plant water requirements are dictated by environmental factors such as climate, soil characteristics, plant canopy size, leaf area, plant development stage, and water supply (Barkataky et al., 2013). The interaction of these factors governs how much water a plant would use for metabolism and physiological needs. Bravdo and Proebsting (1993) note that drip irrigation opens new possibilities for fruit tree growing because it provides means for controlling a few major processes such as water and mineral availability and uptake, and root activity. In order to succeed with drip irrigation, it is important to consider management issues related to: liquid fertigation, frequent irrigation intervals, adoption of the correct type of emitters and spacing, soil type and topography, control of the irrigation depth, and control of the vegetative growth, (Bravdo and Proebsting, 1993).

Principles of Irrigation Scheduling

Drip irrigation can be combined with precision irrigation scheduling, enabling the application of water at a time and rate that are based on a crop's precise water requirement (Madramootoo and Morrison, 2013). With precision irrigation, scheduling is synchronized with weather and soil conditions, and a crop's evapotranspiration rate. Irrigation scheduling relates to water management decisions about: how much water to apply, when to irrigate, and how much to irrigate.

The goal of any irrigation scheduling program is to maintain soil water content close to field capacity in the root zone to optimize crop productivity and minimize leaching and drainage losses. We need to consider the water balance components to accurately determine the plant water requirements. Several researchers have described the major water balance components as inputs [irrigation (IRR), upward flux (UF), effective rainfall (ER)] and outputs/losses [crop evapotraspiration (ET_c), soil moisture changes in root zone (ΔS), surface runoff (R_s), and vertical drainage below the root zone (D)] (Obreza and Pitts, 2002; Jensen, 2007; Fares et al., 2008). Thus the irrigation water requirement becomes:

$$IRR = ET_c + \Delta S - UF - ER$$

The ET_c for any crop is controlled by atmospheric evaporative demand, crop development stage, and available water content. ET_c is calculated from daily reference evapotranspiration (ET_o) as follows:

$$ET_c = ET_o * K_c * K_s$$

Where K_c is the crop coefficient and K_s is the soil water depletion factor (Allen et al., 1998; Morgan et al., 2006).

Effective rainfall (ER) is determined from the following expression:

$$ER = P_{net} - R_s - D$$

Where P_{net} is the net precipitation (mm) reaching the ground surface after canopy interception losses, R_s is the surface runoff (mm) and D is the deep drainage beyond the root zone (mm).

The variation in available soil water storage as an equivalent to soil water depth is expressed as described by (Fares et al., 2000, 2008; Kadyampakeni et al., 2014) as follows:

$$\Delta s = \int_{z_1}^{z_2} \theta(z, t_1)\, dz - \int_{z_1}^{z_2} \theta(z, t_2)\, dz$$

where θ is the soil water content ($cm^3\ cm^{-3}$) at depths z_1 and z_2 (cm).

CROP WATER USE (ETO)

Assuming little or no surface runoff, water applied to the soil surface is 1) retained in the soil, 2) utilized by plants, 3) lost to the atmosphere, or 4) drains below the crop rooting zone. Drainage water may contain substantial quantities of agricultural chemicals and soluble nutrients. Irrigation practices should be aimed at 1) maintaining sufficient water within the crop rooting zone, 2) minimizing pollution of groundwater by leaching, and 3) reducing production costs associated with excessive irrigation, and nutrient and pesticide losses by leaching.

Mills et al., (1999) reported a significant decrease in citrus stomatal conductance after midday. The decrease was most pronounced for south facing exterior leaves and increased with increasing evapotranspirational demand (ET_o). Soil water use from 2 year-old 'Hamlin' orange trees measured at 0.5-hour intervals using weighing lysimeters indicated that water continued to be removed several hours after the midday decrease stomatal conductance. Two seemingly opposing theories place control of soil water uptake at the leaf level via leaf water potential (Slatyer, 1967) or root via root water potential (Tinker and Nye, 2000). The former assumes that leaf water potential exerts control on stomatal conductance regulating transpiration and thus water uptake. The latter speculates that dehydrating roots, due to low soil water content, indirectly control stomatal conductance through the production of chemical compounds that after translocation to the leaves reduce stomatal aperture. Lafolie et al., (1991) measured decreasing leaf water potential with decreased root water potential until midday. After reduced stomatal conductance at midday, leaf water potential increased without a corresponding decrease in root water potential. This was given as evidence that stomatal conductance was not controlled by leaf water potential alone.

FACTORS AFFECTING ET_C

Crop Plant

Citrus are evergreens and therefore require water for transpiration throughout the year. Citrus leaves are thick and waxy, resulting in high cuticular resistance to transpiration (Mills et al., 1999). Koo (1963) and Koo and Sites (1955) stated that water requirements of grapefruit are generally higher than orange or mandarin varieties for trees of equal size. Wiegand and Swanson (1982 a, b, c) and Wiegand et al., (1982) reported that mean daily citrus ET_c at Weslaco, Texas ranged from 2.2 to 3.3 mm for Ruby Red grapefruit and 1.9 to 2.7 mm for Marrs oranges from 5 to 10 years of age.

Under similar climatic conditions, citrus trees are known to have lower transpiration rates compared with other crop plants. Mahrer and Rytwo (1991) reported mean estimated daily ET_c rates for cotton in the Hula Valley of Israel of 5.4 mm when irrigated daily, and 3.96 mm over a 14 day period when not irrigated. Likewise, Starr and Paltineanu (1998) reported that daily ET_c rate for full canopy corn at Beltsville, MD ranged from 3.75 to 5.0 mm prior to rainfall and 5.2 to 8.0 mm after. Lower citrus transpiration rates are related to lower leaf and canopy conductance (Mills et al., 1999).

Tree Size

Large, vigorous, healthy trees require more water than young trees (Tucker et al., 1997). In Florida, large trees at low planting densities (150-180 trees per acre) may use 62.4-94.5 L per day during the winter months and 189-219.2 L per day in July and August (Boman, 1994). Rogers and Bartholic (1976) reported a mean annual ET_c of 1210 mm over an 8-year period from a young orange and grapefruit grove on poorly drained soils near the east coast of Florida. These annual ET_c values ranged from 820 mm early in the study (tree age 2 years) to 1280 mm at end of the study (tree age 10 years). Linear regressions of annual ET_c vs years during the 8 years of the study resulted in significant increase in ET_c. Mean annual ET_c increased at a rate of 19 mm per year or a cumulative increase of approximately 13% over 8 years. Fares and Alva (1999) reported an annual ET_c value of 920 mm for 3-yr old Hamlin orange trees grown on deep sandy soils in central Florida. Koo and Harrison (1965), and Koo and Hunter (1969) reported annual ET_c values of 1170 mm for mature citrus on the same soil series.

Climate

Mean annual ETc for citrus in Florida ranges from 820 to 920 mm (Rogers and Bartholic 1976; Fares and Alva 1999) for young (<5yrs) trees to 1170 to 1280 mm (Koo 1978, Rogers and Bartholic 1976) for mature (10 yrs or more) trees. Annual ETc values reported for mature citrus grown in the lower Rio Grande Valley of Texas are similar to those of Florida and ranged from 1044 to 1232 mm (Wiegand et al., 1982). Hoffman et al., (1982) reported annual ETc values for well-irrigated citrus grown in semi arid Arizona of 1470 mm. Lower ETc rates for Florida (humid) compared with Arizona (semi-arid) have been attributed to lower evaporative demand (Rogers et al., 1983, Fares and Alva 1999).

Soil Characteristics

The supply of water to crops must be based on a clear understanding of the soil water dynamics. Water in excess of field capacity drains through the vadose zone. Eventually, water that is not taken up by plants or evaporated from soil or plant surfaces makes its way into the ground water and contributes to aquifer recharge (Fares and Alva, 1999). Under-tree sprinklers and drip irrigation systems are designed to deliver water at rates low enough to allow infiltration into the soil without contributing to losses by runoff and can be managed to reduce excessive downward drainage through the soil. The required application amount is governed by the soil-water depletion on the irrigation date, irrigation efficiency, and the target soil-water level. The water-balance method is widely accepted for practical use (Prajamwong et al., 1997) and can be generally expressed as

$$d\theta/dt = Irr + Pe + Gw - dSS - RO - ET_c - DP$$

where:

$d\theta/dt$ = change in soil water storage over time
Irr = water added by irrigation
Pe = effective precipitation
Gw = ground-water upflux
dSS = change in surface storage (ponding)
RO = surface runoff
ET_c = actual crop ET, and
DP = deep percolation.

Most of the terms are not independent, for instance, the amount of applied irrigation water will influence the amount of ET_c as well as the amount of DP (Prajamwong et al., 1997).

In standard irrigation practices, water transportation within the soil may be classified in five phases: 1) infiltration during application; 2) redistribution after application ceases; 3) withdrawal by plant roots; 4) evaporation from the soil surface; and 5) drainage of water to deeper soil levels. The primary modes of transport of water in soil are 1) viscous flow through liquid-filled pores, and 2) diffusion of vapor through air-filled pores. In principle, both modes contribute to soil water flow. Liquid flow is the dominant mode in saturated to moist soils (Hagan et al., 1967). Vapor flow does not achieve dominant mode until soils become quite dry, although the presence of a large temperature gradient favors the contribution of this mechanism. For typical soil water situations, both of these modes of transport contribute to a flow rate proportional to potential energy gradients within the soil.

Water is of central importance in the transport of solutes in soils or plants, whether by diffusion or mass flow (Tinker and Nye, 2000). The concept of potential is fundamental to understanding soil water dynamics. Potential is a measure of the energy state of a chemical compound within a particular system, and hence of the ability of a unit amount of the compound to perform work. Difference in potential at different points in a system gives a measure of the tendency of the compound, including water, to move from the region with high potential to the region with the low potential.

Soil water has various forms of potential energy acting upon it, all of which contribute to the total potential. Tinker and Nye (2000) refer to these forms of potential energy as concentration, compression, position in an electrical field, and position in the gravitational field. These same forms of energy are commonly referred to as osmotic, matric, gravitational and pressure potentials, the sum of which is referred to as total water potential(⊖). Thus, soil water moves in response to the difference in water potential over a distance. The first published relationship between water flux and energy gradient was obtained empirically in 1856 by Henry Darcy after a study of the saturated sand filters (Hagan et al., 1967).

Darcy's Law: $\upsilon = -K \, d\varphi / dx$

where:

υ = water flux,
K = hydraulic conductivity constant,
φ = soil water potential, and
x = the distance over which the flux is maintained

The constant of proportionality of Darcy's Law (K) is known as the hydraulic conductivity, and is a funciton of both the properties of the medium and the fluid (Tindall and Kunkel, 1999). In saturated soil, K will be constant so long as the structure of the soil remains stable because the water flow pathways will be unchanged. In unsaturated soil, K varies with the water content (θ), because the latter defines the total cross-section area for water flow, the effective water-filled pore radius, and the effective path length (Tinker and Nye, 2000). A soil with a wide range of pore sizes conducts fluid more rapidly than a soil with small pore sizes (Tindall and Kunkel, 1999). The hydraulic conductivity of soils has a wide range from 10^{-9} cm s^{-1} for clay to 1.0 cm s^{-1} or more for sand. Lower values of K for a clay medium (with smaller pore sizes) are likely due to the drag exerted on the viscous fluid by the walls of the pores. Particles of smaller-sized individual grains (such as clays compared to sands) have a larger surface area that increases the drag on water molecules that flow through the soil, thus, a reduced permeability and hydraulic conductivity.

As water is lost from the soil, the continuity between water-filled pores also decreases. A soil with water-filled volume fractions less than 0.1 ($\theta < 0.1$ cm^3 cm^{-3}) will normally have a very low value for $K(\theta)$ (Tinker and Nye, 2000). The Poiseuille equation states that the flow rate in a tube increases proportionally to the fourth power of its radius, at a constant pressure gradient. Water in larger soil pores will empty first as the soil dries effectively reducing the cross sectional diameter of the soil water pathway. Therefore, pore size and distribution has a large effect on the flow rate

Poiseuille Equation: $f = (\pi r4 / 8n) \, dP/dx$

where:

f = flow rate in a tube,
r = radius,
n = viscosity, and
dP/dx = pressure gradient.

The driving force in soil-water movement is the difference in matric potential, resulting from a difference in soil water content. Richards postulated that Darcy's Law could be extended to unsaturated states by assuming that the hydraulic conductivity (K), as well as the water content, could be treated as non-hysteretic functions of the pressure head or potential (Slatyer, 1967). The matric potential and the water content for a soil are related by the soil-water characteristic curve. By using the slope of the characteristic curve ($d\varphi/d\theta$), the following equation can be obtained based on Darcy's Law and is known as Richard's equation (Tinker and Nye, 2000). In this equation, flow in unsaturated soil can is expressed in terms of the water content gradient and soil water diffusivity (D_θ).

$$\upsilon = -K_\theta \, d\varphi/dx = -K_\theta \, (d\varphi/d\theta)\,(d\theta/dx) = -D_\theta \, (d\theta/dx)$$

where:

υ = water flux,
K_θ = hydraulic conductivity constant,
φ = soil water potential,
θ = soil water content,
$d\varphi/d\theta$ = slope of the soil characteristic curve, and
x = the distance over which the flux is maintained

The term diffusivity (D_θ) is used because the form of equation is the same as that of Fick's law of diffusion (Tinker and Nye, 2000). Furthermore, D_θ is somewhat less convenient than K under conditions of hysteresis because D_θ is discontinuous at each reversal of the direction of potential while K is continuous and virtually hysteresis-free (Hagan et al., 1967). Experimentally, the effect of hysteresis on Richards' equation has usually been ignored by limiting the soil water potential change to either always drying, or always wetting.

"Field capacity" (θ_{fc}) describes the water content held in the soil after excess water has drained to the drier soil layers by redistribution. This equilibrium can be determined in the field by measuring the soil water content as a function of time to determine the value of θ when $d\theta/dt \sim 0$. Hillel (1971) noted that the rate at which $d\theta/dt$ approaches 0 is dependent on θ_i and the depth to which the soil was wetted. The concept of field capacity is useful in the design of field management schemes for approximating the amount of soil water storage. Field capacity can be used as an upper limit value of θ within each soil layer such that any water in excess of θ_{fc} quickly drains to the next deeper soil layer. The soil profile can be thought of as a vertical sequence of reservoirs with the overflow level for each reservoir corresponding to the value of θ_{fc} for that specific soil layer. During irrigation or rainfall the top reservoir flows over to fill the next lower reservoir until no excess water remains to flow into the next reservoir. With a judicious selection of the depth of each soil layer, this simple analog of the soil profile can be easily modeled.

Soil Water Content

Estimated annual ET_c for a deforested area on the Florida ridge reached 680 mm (Sumner, 1996). This ET rate was attributed to periods of low soil water content because the area was not irrigated. Rogers et al., (1983) reported that growth and fruit yield were greater during a 3 year period for treatments maintained at higher soil water content. During the same period of time, annual ET_c averaged 900 and 1210 mm for the lowest and highest soil water content treatments, respectively. Hoffman et al., (1982) reported annual ET_c values of 200 to 500 mm more than Erie et al., (1965) in Arizona. The lower annual ET_c values reported by Erie et al., (1965) were attributed to infrequent irrigation resulting in dry soil surfaces and thus increased resistance to water diffusion to the atmosphere. Smajstrla et al., (1986) reported a reduction in growth and ET_c with increased available soil water depletion of 2-year-old Valencia orange trees grown in drainage lysimeters. Available soil water depletion set points used for irrigation scheduling in this study were 28, 47 and 58%. It was concluded that tree stress occurred at the highest depletion value due to the reduced ability of the soil to transport water to the roots because of reduction in hydraulic conductivity. Fares and Alva (1999) calculated daily ET_c for 3-year-old Hamlin orange trees on deep sandy soil in central Florida. Estimated daily ET_c values decreased with time after each rainfall or irrigation.

Water Table

Obreza and Admire (1985) concluded that shallow water tables in the flatwoods soils could significantly augment water available for root uptake. Graser and Allen (1987) suggested that water-table management by controlling water table depth in the winter and spring could help decrease the need for supplemental irrigation during the dry season. Boman (1994) used drainage lysimeters in which he maintained a water table at 0.61, 0.76, and 0.91 m to measure the effects of water table on ET_c, growth, yield and fruit quality of 5-year-old Valencia trees. No significant differences were found at the three water table depths.

Soil Shading

Castel et al., (1987) estimated soil surface evaporation by comparing water loss from weighing lysimeters in which the soil was covered by plastic with lysimeters that remained uncovered. Mean estimated evaporation was reported as 0.775 mm of an estimated potential ET of 4.25 mm, which was greater than 18%. Castel and Buj (1992) reported that the percentage of ground shaded by young Clementine trees increased from 10 to 25% during a 4-year period from 10% to 25%. Evapotranspiration increased by 33% during the same time period. This increase was attributed to the increasing water use by the trees and reduced soil surface evaporation.

Grass and Weed Growth

Smajstrla et al., (1986) used field drainage lysimeters to determine the effect of grass cover on the growth and ET_c of 2-year-old Valencia orange trees. Automated covers were installed to cover the lysimeters during rainfall. Soil within the lysimeters was maintained bare or covered completely with bahiagrass. The bare soil lysimeters consistently had the greatest monthly ET_c. Measured annual ET_c ranged from 1331 to 900 mm for grass covers lysimeters and 912 to 441 mm for those with bare soil surfaces. Total ET_c increased 46 to 105% per year due to soil grass cover. These results were similar to those reported by Stewart et al., (1969) using non-weighing lysimeters. In their study estimated annual bare soil evaporation and 2/3 sod cover ET_c averaged 68 and 92% of full sod cover, respectively. Tucker et al., (1997) reported reduced soil water use from non-irrigated middles between rows of mature citrus by limiting the height of weed growth by chemical mowing.

Crop Coefficient (K_c)

The crop coefficient (K_c) incorporates crop characteristics and averaged effects of evaporation from the soil (Allen et al., 1998) and is proportional to atmospheric demand and plant development stage (Fares et al., 2008). The K_c is used for normal irrigation planning and management purposes, for the development of basic irrigation schedules, and for most hydrologic water balance studies. Allen et al., (1998) outlined the steps for calculating ET_c as follows: 1) identifying the crop growth stages, determining their lengths, and selecting the corresponding K_c coefficients; 2) adjusting the selected K_c coefficients for frequency of wetting or climatic conditions during the stage; 3) constructing the crop coefficient curve (allowing one to determine Kc values for any period during the growing period); and 4) calculating ET_c as the product of ET_o and K_c. Doorenbos and Pruitt (1975) provides general lengths for the four distinct growth stages and the total growing period as follows 1) initial, 2) crop development, 3) mid-season, and 4) late-season stages. The respective durations of these stages in a 365-day calendar year are as follows: initial stage (60 days), development stage (90 days), mid-season stage (120 days), and late season stage (95 days).

The citrus crop coefficients have been estimated over the years for Mediterranean and sub-humid environments (Table 1). These values are close to those calculated for young and mature citrus which range from 0.6 in winter to 1.2 in summer (Martin et al., 1997; Fares and Alva, 1999; Morgan et al., 2006; Fares et al., 2008). Morgan et al., (2006), in particular, established a quadratic function for predicting K_c for microsprinkler-irrigated mature citrus as a function of day of the year (Figure 1) while accounting for available soil water depletion for sandy soils of central Florida (Figure 2). De Azevedo et al., (2008) found improved water management with drip irrigation through manipulation of irrigation cycles and the grapevine K_c was adequately described as a function of days by polynomial models ($R^2 \approx 0.71 - 0.90$).

Table 1. Approximate time-averaged crop coefficients (K_c) and mean maximum plant heights for non-stressed, well-managed crops in subhumid climates for use with the FAO Penman-Monteith ET_o (adapted from Allen et al., 1998)

Crop	K_c initial	K_c mid-season	K_c end season	Maximum crop height (m)
Citrus, no ground cover				
70% canopy	0.70	0.65	0.70	4
50% canopy	0.65	0.60	0.65	3
20% canopy	0.50	0.45	0.55	2
Citrus, with ground cover				
70% canopy	0.75	0.70	0.75	4
50% canopy	0.80	0.80	0.80	3
20% canopy	0.85	0.85	0.85	2

Rogers et al., (1983) reported monthly measured ET_c to calculated ET_o ratio values using the mean of four methods of estimating ET_o (Penman, Blaney-Criddle, Jensen-Haise, and Class A pan). The resulting monthly ratios range from 0.9 in January to 1.11 in June. Crop coefficient (K_c) values reported by Doorenbos and Pruitt (1977) after adjustments for humid conditions ranged from 0.9 in March though December to 0.95 in January and February. Castel et al., (1987) estimated monthly K_c for drip-irrigated mature navels grown in Valencia, Spain. Their K_c values were calculated from mean daily ET_c estimated from weekly ET values determined by neutron probe measurements. Values ranged from an mean of 0.71 from January – July to 0.90 from August – December. Castel and Buj (1992) suggested these values differed from those reported for Florida due to the lower evaporative demand of the humid Eastern coast of Spain, which has an mean annual ET_o of 1166 mm compared with 1400 mm in Florida.

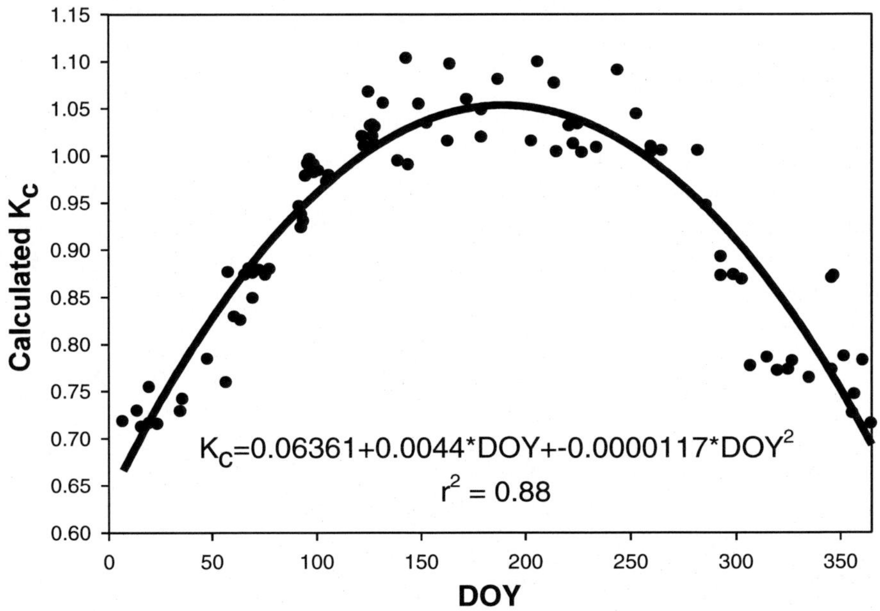

Figure 1. Generalized crop coefficient curve for the single crop coefficient approach (from Morgan et al., 2006).

Calculated K_c values for 3-year-old Hamlin trees grown on sandy soil in central Florida ranged from approximately 1.05 in November - March to 0.85 in May – August (Fares and Alva 1999). Boman (1994) calculated K_c values for 5-year-old Valencia orange trees grown in non-weighing lysimeters with water tables maintained at 0.6, 0.75, or 0.9 meters from the soil surface. Calculated K_c values were at a minimum of 0.6 during December – February and peaked at 1.1 in June and July. Martin et al., (1997) estimated mean daily ET_c values for 7-year-old "Redblush" grapefruit in Arizona from soil water content data collected at 1 to 2 week intervals. Monthly K_c values were calculated by comparing these estimated daily values with mean daily ET_o for the same period. The resulting K_c ranged from a low of 0.55 – 0.6 in December and January to a high of 1.1 – 1.2 in July.

Soil Water Depletion Coefficient

According to Allen et al., (1998), water depletion coefficient is defined as the effect of soil water reduction on ET_c by reducing the value of K_c. This is calculated by multiplying the K_c of a given crop by the soil water depletion coefficient (K_s) for a given soil water content. Water stress increases as soil water is extracted by evapotranspiration.

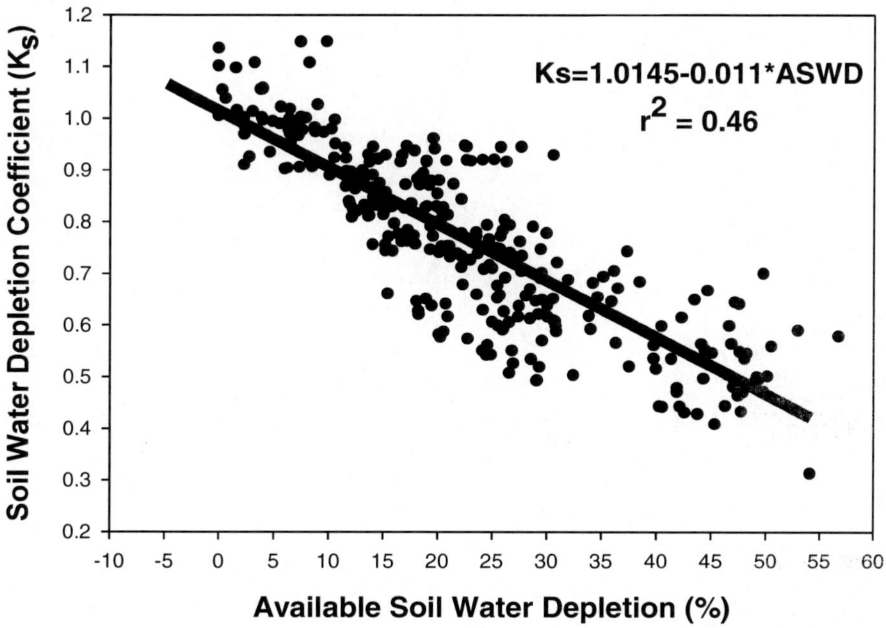

Figure 2. Reduction in water uptake with decrease estimated by the single crop coefficient approach (from Morgan et al., 2006).

Available soil water (ASW) is defined as the difference between drained upper limit (field capacity) and the drained lower limit or wilting point. However, the energy expenditure required to extract residual soil water increases as soil water content decreases. Likewise, resistance to water flow increases as residual soil water decreases reducing water flux to the root boundary. Therefore, crop water uptake is reduced well before wilting point is reached (Allen et al., 1998). At field capacity, roots can absorb water fast enough to supply the ET_c demand of the atmosphere. However, water becomes more strongly bound to the soil matrix and is more difficult to extract as soil water content decreases. When soil water content drops below a threshold value, soil water can no longer be transported quickly enough to the roots to supply the transpiration demand of the crop. The fraction of ASW above this threshold is known as readily available water (RAW). For most crops grown on medium and fine textured soils RAW is as much as 30 to 50% of ASW (Allen et al., 1998). When root zone depletion exceeds this threshold ET is reduced to less than potential crop ET_c and stress begins.

Morgan et al., (2006) developed a relationship for K_s under field conditions with sandy soils (sand content >95%) by correlating measured water uptake with Penman Monteth estimated ET_o (Figure 2). Soil water use decreased with soil water content, resulting in a K_s value that dropped from 1.0 at ≤ 1% ASWD to 0.5 at 50% ASWD. With few exceptions, daily soil water uptake per unit root length density was similar for all soil layers. The best correlation between daily water use and soil water content was found in the soil volume containing the highest root length density.

CITRUS ROOT GROWTH DYNAMICS

Citrus trees are productive and grow well on central Florida deep sandy soils. Trees on many rootstocks are well adapted to such soils and produce extensive, deep root systems as demonstrated in several Florida studies. In some instances, tree size and yield appear to be related to root system characteristics (Castle and Kresdorn, 1975).

Citrus fibrous roots are commonly defined as those roots < 2 mm in diameter. Their dry mass is a small part of the total root system, but their composited length far exceeds that of the woody roots (> 2 mm in diameter). These fine roots are considered to be the "functional" part of the root system because of their role in water and nutrient uptake. There is some variation among rootstocks in the morphology of the fibrous roots (Castle and Youtsey, 1977).

Some rootstocks like trifoliate orange *[Poncirus trifoliata* (L.) Raf.] produce higher specific root length, or length/unit mass (Eissenstat, 1991). Fibrous roots are also the most vulnerable part of the root system. Their development, function, and longevity are strongly influenced by soil characteristics, environmental changes, crop species, crop growth stage, and cultural practices.

FACTORS AFFECTING ROOT DISTRIBUTION AND ROOT DENSITY

Soil Characteristics

The distribution of roots is modified by the physical and chemical properties of the soil profile (Hillel, 1971). Widespread root development and high fibrous root concentrations were observed in deep soils of sand texture where there were virtually no impediments to root growth provided that water and nutrients are non-limiting to growth (Ford 1952; 1953a&b; 1954a&b; 1959; 1964; 1972); Ford et al., (1954; 1957). Increased tree size and yield have been related to root system depth and fibrous root mass. The depth of rooting of 'Orlando' tangelo trees on 10 rootstocks growing in deep, sandy soil was correlated with tree height (Castle and Krezdorn, 1975). Although fibrous root distribution was affected by tree height, total fibrous root dry mass as measured at the canopy dripline was not correlated with tree height. Ford (1954a; 1964; 1968; 1969; 1972) conducted many studies of citrus trees in poorly drained Spodosols of Florida and concluded that tree size was closely related to fibrous root density. Extensive lateral root development occurred on soils with loamy or clay texture (Boswell et al., 1975; Kaufinann et al., 1972). In these studies the root systems were shallower with few roots found deeper than about 50 to 70 cm in the soil profile (Adriance and Hampton, 1949; Boswell et al., 1975; Cahoon et al., 1956; 1959; 1961; Kimball et al., 1950; Mikhail and El-Zeftawi, 1978). Furthermore, the changes in fibrous root distribution with depth were more gradual compared with sandy soils, and the fibrous root concentrations were lower (Bielorai, 1977). The lower natural fertility (Carlisle et al., 1989), and excessive drainage of sandy soils resulted in higher shoot:root ratios such that fibrous root dry mass densities tended to be lower in sandy soils (Castle, 1978).

Under flatwood conditions where the soil is drained and bedded, virtually all the root mass occurs within 45 cm of the soil surface (Calvert et al., 1967; 1977; Ford, 1954a; Ford, 1972; Reitz and Long, 1955). The quantity of fibrous roots decreases with depth and lateral distance from the trunk. Elezaby (1989) reported the lateral fibrous root distribution to a depth of 180 cm of a 10-year-old 'Valencia' tree *[Citrus sinensis* (L.) Osb.] on 'Volkamer' lemon *(C. volkameriana* Ten. and Pasq.) grown on a soil with a deep sand profile and spaced at 4.5 m x 6.0 m as: 9% of the fibrous roots between 0 cm and 60 cm from the trunk, 31% between 120 cm and 180 cm, and 21% between 240 cm and 300 cm.

The vertical distribution was 42% of the fibrous roots were between 0 cm and 30 cm from the soil surface, and 14% or less at each 30-cm depth increment to 180 cm. Those data are similar to dry mass densities reported in other Florida studies (Castle, 1978; 1980). In another research report, data were given as root length densities and ranged from 530 cm m^{-3} for 'Swingle' citrumelo roots *[C. paradisi* Macf. x *P. trjfoliata* (L.) Raf] to 2,020 cm m^{-3} for trifoliate orange (Eissenstat, 1991).

Climatic Effects

Root distribution was studied in 22 mature navel or 'Valencia' orange groves in California. In this study, 50% were low yielding while the remaining 50% were high producing. Fibrous root fresh mass was measured to a depth of 90 cm under the canopy and between rows (Cahoon et al., 1956). Yield was not related to the under-canopy root quantities, but was correlated with the root quantities measured between adjacent rows where soil water contents were typically lower most of the year.

Rootstocks

Some citrus roots have been found as deep as 7 m (Ford, 1954b*),* and in one instance, roots of mature trees on rough lemon rootstock were discovered 14 m from the tree trunk (Ford, 1970). Castle and Krezdorn (1975) described two general types of root systems, the first being characterized as "extensive" featuring extensive lateral and vertical development, and second as "intensive" with less extensive root expansion and higher fibrous root concentrations mainly confined to the upper soil layers. Trees on rough lemon, 'Volkamer' lemon and 'Palestine' sweet lime (*C. limettioides* Tan.) typified the extensive type of root system where 50% of the fibrous roots occurred below 70 cm in the soil with wider spreading lateral development. Examples of the intensive type were 'Rusk' citrange and trifoliate orange, where few fibrous roots were found below 70 cm, and the root system was less developed laterally. Some rootstocks like sour orange and 'Cleopatra' mandarin were classified as intermediate. Trees on 'Cleopatra' mandarin had a highly developed lateral root system at the surface, but few fibrous roots below the surface. Menocal-Barberena (2000) found no statistically significant differences in vertical or horizontal fibrous root distribution of 'Hamlin' orange on four different rootstocks. These rootstocks were sour orange,

'Cleopatra' mandarin, 'Swingle' citrulemo and 'Carrizo' citrange. Vertical and horizontal root distribution was similar to other studies with about 40% of the fibrous roots in the top 30 cm and 9-14% at each of the remaining 30 cm depth increments to 180 cm. Few roots were found below 180 cm.

Tree Spacing and Density

Due primarily to Florida's rainy season, roots of trees at commercial spacings rapidly occupy the volume of soil outside the irrigated zone. After canopy closure, they extend into the rootzone of adjacent trees. Elezaby (1989) reported that fibrous root concentration in the 0-30 cm zone increased from 450 g m^{-3} to 1,000 g m^{-3} between trees when the in-row distance decreased from 4.5 m to 2.5 m. The increased root concentrations in this study were concluded to be the result of overlapping root systems. Trees at the closest spacings showed root concentration increases to depths of 150 cm (Elezaby, 1989).

Fertilization

Increases in fertilizer N can increase root growth to a considerable depth, but the largest effects generally occurred near the surface (Ford, 1953b; Ford et al., 1957; Smith, 1956; Smith, 1965).

Irrigation

Irrigation method and scheduling has been shown to change the distribution and/or concentration of citrus fibrous roots. In a California study of trees receiving differential irrigation treatments, yield was not correlated with fibrous root density (Cahoon et al., 1964) because trees on the low irrigation rate declined in yield while maintaining root quantities similar to those of the trees on the higher irrigation rate. It was concluded that soil water content was the single most important factor influencing citrus root systems.

In a Florida study, root weight densities were determined under the tree canopy, at the dripline, and in the row middles to a depth of 180 cm for 'Hamlin' orange trees on 'Swingle' citrumelo and 'Carrizo' citrange rootstocks (Menocal-Barberena, 2000). Trees receiving irrigation at a rate of 40 cm yr^{-1} had significantly higher densities than trees receiving 250 cm yr^{-1}. The differences were on the order of 1.3 to 2.3 times greater for the 40 cm yr^{-1} treatment at all depths.

Canopy Reduction

Hedging, the annual removal of excess vegetative growth, has become a common method of canopy size control for closely planted citrus trees. Eissenstat and Duncan (1992) found that within 30 days of canopy reduction, 20% of the fibrous roots between the 9 cm and 35 cm depths were apparently dead. Root length density of these trees recovered within 63 days of canopy reduction. This relatively short-term reduction in fibrous root density adversely affected yield because of fruit abortion.

COLD ACCLIMATION AND FREEZE PROTECTION

During winter, water uptake by citrus trees also declines due to the resistance exerted by the roots resulting in low evapotranspiration rates. Hilgeman et al., (1969) found that internal water stress was at a minimum in March and April, however, in early May there quickly followed a period of high transpiration and large internal water deficits. The authors also found cooler cloudy conditions lowered transpiration in August and transpiration increased again on warmer days in September and early October, only to decrease once more with the onset of winter due to low temperatures and low vapor deficits.

Low temperature (LT) is one of the most important abiotic factors limiting growth, productivity and distribution of plants (Boyer, 1982; Sakai and Larcher, 1987). LT decreases biosynthetic activity of plants, inhibits the normal function of physiological processes and may cause permanent injuries, leading to death of the plant. Freezes cause ice to form in the apoplast (Meryman, 1956, 1966; Yelenosky, 1985, 1996). When soil moisture is not limiting, unacclimated citrus have fully turgid apoplasts such that freezing cause apoplastic ice to penetrate the symplast directly (Young and Mann, 1974). During extracellular freezing, ice first forms in the dilute apoplastic solution and a water potential gradient is established between the extracellular ice crystal and the intracellular liquid water. The lower water potential of ice as compared to that of liquid cellular water at the same temperature will cause liquid water to move from the cell to the extracellular ice (Levitt, 1980). This process, depending on the temperature and concentration of the cell sap imposes a dehydration stress on the cell. When the cellular water potential of the partially dehydrated cell equals that of the extracellular ice, an equilibrium is established and further dehydration will not occur provided the temperature remains constant (Rajashekar and Burke, 1982). If the temperature further declines or increases, water will flow out of or back into the cell, respectively. During such equilibrium freezing a tissue can behave either as an ideal solution or as a non ideal solution (Rajashekar and Burke, 1982). In non ideal equilibrium freezing, negative wall pressure is believed to reduce the degree of cell dehydration and solute concentration (Rajashekar and Burke, 1982). Cold tolerance of citrus is promoted by cold temperature (Ebel et al., 2005; Yelenosky et al., 1984; Yelenosky, 1978, 1985, 1996; Young, 1961) which is accompanied by plant dehydration (Yelonosky, 1982; Yelenosky, 1985), an important mechanism that enhances survival during freeze events.

In order to understand how plants behave in cold temperature, it is necessary to know the water movement within the plant. Water movement in soil plant atmosphere continuum can be described using an analogy of Ohm's law (Landsberg and Jones, 1981; Van den Honert, 1948), which describes flow as being proportional to driving force (the water potential gradient) and inversely proportional to the resistance in the flow path:

$$T = (p_{vs} - p_{va}) / (r_{vs} - r_{va}) = (\Psi_{soil} - \Psi_{stem}) / (R_{root} + R_{soil})$$

where, T is transpiration, p_{vs} is vapor pressure at the leaf surface, p_{va} is the vapor pressure of the atmosphere, r_{vs} is stomatal resistance and the inverse is stomatal conductance, r_{va} is boundary layer resistance, Ψ_{soil} is soil water potential, Ψ_{stem} is stem water potential, R_{root} is root resistance and R_{soil} is resistance exerted by soil against water movement from soil to the root surface.

This can be rearranged to solve for R_{root}:

$$R_{root} = ((\Psi_{soil} - \Psi_{stem}) / T) - R_{soil}$$

It is assumed that Ψ_{soil} for soil at water holding capacity is about – 0.03 MPa such that assuming that it is zero in our study introduces a relatively small error. Also, it is assumed that R_{soil} was zero, which is a reasonable assumption for well watered soils. These assumptions simplify equation 2 to:

$$R_{root} = - \Psi_{stem} / T$$

Ψ_{stem} is measured and T is determined by:

$$T = (PW_o - PW_t) / A_{leaf} / t$$

where, PW is pot weight (g) measured 2 hrs watering (PW_o) and about 4 hrs later (PW_t), A_{leaf} is total leaf area of the plant ($m^{2)}$, and t is the time elapsed between weight measurement (s). To determine R_{root} in units of MPa s/m, units that are the most commonly reported in the literature for liquid resistance, T was multiplied by 1×10^{-6} m^3 g^{-1} to convert transpiration units from weight ($g\ m^{-2}\ s^{-1}$) to volume ($m^3\ m^{-2}\ s^{-1}$).

Transpiration is driven by evaporative demand ($p_{vs} - p_{va}$), which is controlled by stomata at the leaf-air interface and accelerated by wind which reduces $r_{va.}$ As the VPD (vapor pressure deficit) increases, stomatal conductance of citrus tends to decrease. Citrus having high supply resistance, stomatal sensitivity to VPD may have influence on maintaining favorable leaf water relations (Kriedemann and Barrs, 1981). Hall et al., (1975) found higher stomatal conductance in controlled environment when VPD was low, but found relatively lower values at other times. Stem Ψ is the result of whole plant transpiration, and soil and root/soil hydraulic conductivity (Chone et al., 2001). Evaporation of water from the stomatal cavities results in development of a plant water deficit due to the decline of Ψ_{stem} and when Ψ_{stem} gets lower than Ψ_{soil}, water is drawn from soil into the plant. R_{root} represents about 2/3 of total plant hydraulic resistance (Landsberg and Jones, 1981; Passioura, 1988) and hence is one of the most important factors that control soil-plant-atmosphere continuum. During cold acclimation, dehydration of citrus is partly caused by the increase in root resistance which inhibits water uptake (Kadoya et al., 1981; Kramer, 1969; Nielson, 1974; Syvertsen et al., 1983). With the exception from the usual case of stomatal aperture, where maximum dehydration is brought when all the stomata are open; cold acclimated citrus experience dehydration even though the stomata are closed. This may be because of higher root resistance, (Kadoya et al., 1981) which inhibits the water uptake by the roots resulting in dehydration.

Conclusion

The paper presented a framework and a broad perspective for improved citrus water management. Citrus physiological responses and root distribution patterns have a bearing on citrus water management. Improvements in managing these responses, particularly through a sustained use of precision irrigation management tools, will be critical in achieving improved citrus production globally.

References

Allen, R.G., L.S. Pereira, D. Raes, and M. Smith. 1998. Crop evapotranspiration – guidelines for computing crop water requirements – FAO Irrigation and drainage paper 56. *Food and Agriculture Organization of the United Nations,* Rome, Rome, 300 p.

Barkataky, S., K.T. Morgan and R.C. Ebel. 2013. Plant water requirement of 'Hamlin' sweet orange in cold temperature conditions. *Irrigation Science* 31:431–443.

Bielorai, H. 1977. The effect of drip and sprinkler irrigation on grapefruit yield, water use and soil salinity. *Proc. Int. Soc. Citriculture* 1:99-103.

Boman, B. J. 1994. Evapotranspiration by young Florida flatwoods citrus trees. *J. Irrigation and Drainage Engineering* 120:80-88.

Boswell, S.B., C.D. McCarty, and L.N. Lewis. 1975. Tree density affects large-root distribution of 'Washington' Navel orange trees. *HortScience* 10:593-595.

Boyer JS (1982) Plant productivity and environment. *Science* 218: 443-448.

Bravdo, B. and E.L. Proebsting. 1993. Use of drip irrigation in orchards. *HortTechnology* 3(1):44-49.

Cahoon, G.A., R.B. Harding, and D.B. Miller. 1956. Declining citrus root systems. *Calif. Agric.* 10(9):3-12.

Cahoon, G.A., E.S. Morton, W.W. Jones, and M.J. Garber. 1959. Effects of various types of nitrogen fertilizers on root density and distribution as related to water infiltration and fruit yields of 'Washington' Navel oranges in a long-term fertilizer experiment. *Proc. Amer. Soc. Hort. Sci.* 74:289-299.

Cahoon, G.A., M.R. Huberty, and M.J. Garber. 1961. Irrigation frequency effects on citrus root distribution and density. *Proc. Amer. Soc. Hort. Sci.* 77:167-172.

Castel, J.R., I. Bautista, C. Ramos, and G. Cruz. 1987. Evapotranspiration and irrigation efficiency of mature orange orchards in 'Valencia' (Span). *Irrigation and Drainage Systems* 3:205-217.

Calvert, D.V., RCJ. Koo, and H.W. Ford. 1967. Flood irrigation studies with citrus. *Proc. Fla. State Hort. Soc.* 80:79-85.

Calvert, D.V., H.W. Ford, E.H. Stewart, and F.G. Martin. 1977. Growth response of twelve citrus rootstock-scion combinations on a Spodosol modified by deep tillage and profile drainage. *Proc. Intl. Soc. Citriculture* 1:79-84.

Carlisle, V.W., F. Sodek, M.E. Collins, L.C. Hammond, and W.G. Harris. 1989. Characterization data for selected Florida soils. University of Florida-Soil Science Department, Soil Characterization Laboratory, Institute of Food and Agricultural

Sciences in cooperation with U.S. Department of Agriculture-Soil Conservation Service. *Soil Science Research Report* No. 89-1. 307 p.

Castel, J.R., and A Buj. 1992. Growth and evapotranspiration of young, drip-irrigated Clementine trees. *Proc. Int. Citriculture*, 651-656.

Castle, W.S. 1978. Citrus root systems: Their structure, function, growth, and relationship to tree performance. *Proc. Int. Soc. Citriculture* 1:62-69.

Castle, W.S. 1980. Fibrous root distribution of 'Pineapple' orange trees on rough lemon rootstock at three tree spacings. *J. Amer. Soc. Hort. Sci.* 105(3):478-480.

Castle, W.S. and A.H. Krezdorn. 1975. Effect of citrus rootstocks on root distribution and leaf mineral content of 'Orlando' tangelo tree. *J. Amer. Soc. Hort. Sci.* 100(1):1-4.

Castle, W.S. and C.O. Youtsey. 1977. Root system characteristics of citrus nursery trees. *Proc. Fla. State Hort. Soc.* 90:39-44.

Chone, X., C.van Leeuwen, D. Dubourdieu and J. P. Gaudillere. 2001. *Annals of Botany* 87: 477-483.

de Azevedo, P.V., J.M. Soares, V. P. R. da Silva, B.B. da Silva, and T. Nascimento. 2008. Evapotranspiration of ''Superior'' grapevines under intermittent irrigation. *Agricultural Water Management* 95:301–308.

Doorenbos, J. and Pruitt, W.O. 1975. Guidelines for predicting crop water requirements, Irrigation and Drainage Paper 24, *Food and Agriculture Organization of the United Nations, Rome*, 179 p.

Ebel, R. C., B. L. Campbell, M.L. Nesbitt, W. A. Dozier, J. K. Lindsey, and B. S. Wilkins. 2005. A temperature index model to estimate long-term freeze risk of Satsuma mandarins grown on the northern coast of the Gulf of Mexico. *J. Amer. Soc. Hort. Sci.* 130: 500-507.

Eissenstat, D.M. 1991. On the relationship between specific root length and the rate of root proliferation: a field study using citrus rootstocks. *New Phytol.* 118:63-68.

Eissenstat, D.M. and L.W. Duncan. 1992. Root growth and carbohydrate responses in bearing citrus trees following partial canopy removal. *Tree Physiol.* 10:245-257.

Elezaby, A.A. 1989. Physiological and biological studies on root systems of some citrus stocks. Ph.D. Dissertation. Cairo University, Egypt.

Erie, L.J., O.F. French, and K. Harris. 1965. Consumptive use of water by crops in Arizona. *University of Arizona Agri. Exp. Sta. Tech. Bul. No.* 169, 44 pp.

Fares, A., and A.K. Alva. 1999. Estimation of citrus evapotranspiration by soil water mass balance. *Soil Science* 164:302–310.

Fares, A., A.K. Alva, P. Nkedi-Kizza, and M.A. Elrashidi. 2000. Estimation of soil hydraulic properties of a sandy soil using capacitance probes and Guelph permeameter. *Soil Science* 165:768–777.

Fares, A., A. Dogan, F. Abbas, L. Parsons, T. Obreza, and K. Morgan. 2008. Water balance components in a mature citrus orchard. *Soil Science Society of America Journal* 72:578-585.

Ford, H.W. 1952. The distribution of feeder roots of orange and grapefruit trees on rough lemon rootstock. *Citrus Industry Magazine* 14(1 l):22-23.

Ford, H.W. 1953a. Effect of spreading decline disease on the distribution of feeder roots of orange and grapefruit trees on rough lemon rootstock. *Proc. Amer. Soc. Hort. Sci.* 61:68-72.

Ford, H.W. 1953b. Root distribution of chlorotic and iron-chelate-treated citrus trees. *Proc. Fla. State Hort. Soc.* 66:22-26.

Ford, H.W. 1954a. Root distribution in relation to the water table. *Proc. Fin. State Hort. Soc.* 67:30-33.

Ford, H.W. 1954b. The influence of rootstock and tree age on root distribution of citrus. *Proc. Amer. Soc. Hort. Sci.* 63:137-142.

Ford, H.W. 1959. Growth and root distribution of orange trees on two different rootstocks as influenced by depth to subsoil clay. *Proc. Amer. Soc. Hort. Sci.* 74:313-321.

Ford, H.W. 1964. The effect of rootstock, soil and soil pH on citrus root growth in soils subject to flooding. *Proc. Fin. State Hort. Soc.* 77:41-45.

Ford, H.W. 1968. Fluctuations of the water table in drained flatwoods groves. *Proc. Fla. State Hort. Soc.* 81:75-79.

Ford, H. W. 1969. Water management of wetland citrus in Florida. Proc. First Int. *Citrus Symp.* 3:1759-1770.

Ford, H.W. 1970. Problems in using 'Milam' rootstock as a biological barrier. *Proc. Fla. State Hort. Soc.* 83:84-86.

Ford, H.W. 1972. Eight years of root injury from water table fluctuations. *Proc. Fla. State Hort. Sci.* 85:65-68.

Ford, H.W., I. Stewart, and C.D. Leonard. 1954. The effect of iron chelate on root development of citrus. *Proc. Amer. Soc. Hort. Sci.* 63:81-87.

Ford, H.W., W. Reuther, P.F. Smith. 1957. Effect of nitrogen on root development of 'Valencia' orange trees. *Proc. Amer. Soc. Hort. Sci.* 70: 237-244.

Graser, E.A., and L.H. Allen, Jr. 1987. Water relations of 7-year-old containerized citrus trees under drought and flooding stress. *Proc. Fla. State Hort. Soc.* 100:126-136.

Hagan, R. M., H. R. Haise, and T.W. Edminster (eds). 1967. Irrigation of Agricultural Lands. *Am. Soc. of Agron.* Number 11.

Hall, A. E., Camacho-B., S. E., and Kaufman, M. R. 1975. Regulation of water loss by citrus leaves. *Physiol. Plant.* 33, 62-65.

Hilgeman, R. H., Ehrler, W. L., Everling, C. E., and Sharp, F. O. 1969. Apperant transpiration and internal water stress in Valencia oranges as affected by soil water, season and climate. *Proc. Int. Citrus Symp.* 1st, 1968 Vol. 3, pp 1713-1723.

Hillel, D. 1971. Soil and water physical principles and processes. Academic Press, New York, New York.

Hillel, D. 1998. Environmental soil physics. Academic Press. New York, New York.

Hoffman, G.J., J.D. Oster, and W.J. Alves. 1982. Evapotranspiration of mature orange trees under drip irrigation in an arid climate. *Trans. Am. Soc. Agric. Engr.* 25:992-996.

Jensen, M.E. 2007. Beyond irrigation efficiency. *Irrigation Science* 25:233-245.

Kadyampakeni, D.M., K.T. Morgan, A.W. Schumann, P. Nkedi-Kizza, and T.A. Obreza. 2014. Water use in drip- and microsprinkler-irrigated citrus trees. *Soil Science Society of America Journal* 78:1351–1361.

Kadoya, K., F. Mizutani, and A. Hino. 1981. Effects of root environmental conditions on stomatal behavior and translocation of photosynthates. *Proc. Intl. Intl. Soc. Citriculture.* 2: 498-501.

Kimball, M.H., A. Wallace, and R.T. Mueller. 1950. Changes in soil and citrus root characteristics with non-tillage. *California Citrograph* 35:432-433.

Kramer, P. J. 1969. Plant and soil water relationships. McGrwa-Hill, Inc. p. 482.

Kriedemann, P. E. and H. D. Barrs. 1981. Citrus Orchards. Chapter 5. *Water deficits and plant growth.* VI: 325-417.

Koo, R.C.J. 1963. Effects of frequency of irrigation on yield of orange and grapefruit. *Proc. Fla. State Hort. Soc.* 74:1-5.

Koo, R.C.J. 1978. Response of densely planted 'Hamlin' orange on two rootstocks to low volume irrigation. *Proc. Fla. State Hort. Soc.* 91:8-10.

Koo, R.C.J., and J.W. Sites. 1955. Results of research and response of citrus to supplemental irrigation. *Proc. Soil Sci. Soc. Fla. 15*:180-190.

Koo, R.C.J., and D.S. Harrison. 1965. Summary of irrigation research in Florida. *Florida Agricultural Ext. Service Memo*. Report 65-8.

Koo, R.C.J., and G.T. Hurner, Jr. 1969. Irrigation requirements of citrus grown on Lakewood fine sand. *Proc. Fla. State Hort. Soc.* 82:69-72.

Lafolie, F., L. Bruckler, and F. Tardieu. 1991. Modeling root water potential and soil- root water transport: I. model presentation. *Soil Sci. Soc. Am. J.* 55:1203-1212.

Landsberg, J. J. and H. G. Jones. 1981. Water deficits and plant growth. Chapter 6, Apple Orchards, Academic Press, New York, NY.

Levitt, J., 1980. Responses of Plants to Environmental Stresses. Vol.1. *Chilling, Freezing, and High Temperature Stresses*. Academic Press, New York.

Madramootoo, C.A. and J. Morrison. 2013. Advances and challenges with micro-irrigation. *Irrigation and Drainage* 62:255–261.

Mahrer, Y., and G. Rytwo. 1991. Modeling and measuring evapotranspiration in a daily drip irrigated cotton field. *Irrig. Sci.* (1991) 12:13-20.

Martin, E.C., A.K. Hla, P.M. Waller, and D.C. Slack. 1997. Heat unit-based crop coeffi cient for grapefruit trees. *Applied Engineering in Agriculture* 13:485–489.

Menocal-Barberena, 2000. Effect of rootstock on root distribution of citrus. M.S. Thesis. University of Florida, Gainesville, Fl.

Meryman, H. T. 1956. Mechanics of freezing in living cells and tissues. *Sci.* 124: 515-521.

Meryman, H. T. 1966. Review of biological freezing, p.2-114. In: H. T. Meryman 9ed.) Cryobiology, Academic Press, New York.

Mikhail, E.H. and B.M. El-Zeflawi. 1978. Effect of soil types and rootstocks on root distribution and leaf composition of citrus trees. *Proc. Int. Soc. Citriculture* 1: 214-216.

Mills, T.M., K.T. Morgan, and L.R. Parsons. 1999. Canopy position and leaf age affect stomatal response and water use in citrus. *J. Crop Production* 2(2):163-179.

Morgan, K., T. Obreza, J. Scholberg, L. Parsons, and T. Wheaton. 2006. Citrus water uptake dynamics on a sandy Florida Entisol. *Soil Science Society of America Journal* 70:90-97.

Nielsen, K. F. 1974. Roots and root temperatures. In: E. W. Carson (ed.) The Plant Root Environment, Carson Univ. Press, Virginia.

Obreza, T.A., and K.E. Admire. 1985. Shallow water table fluctuations in response to rainfall, irrigations, and evapotranspiration in flatwood citrus. *Proc. Fla. State Hort. Soc.* 68:24-29.

Obreza, T., and D. Pitts. 2002. Effective rainfall in poorly drained microirrigated citrus orchards. *Soil Science Society of America Journal* 66:212-221.

Pessioura, J. B. 1988. Water transport in and to roots. *Ann. Rev. Rev. Plant Physiol. Plant Mol. Biol. 39*: 245-265.

Prajamwong, S., G. P. Merkley, and R. G. Allen. 1997. Decision Support Model for Irrigation Water Management. *J. Irrig. And Drain. Engrg*., ASAE, 122(2):106-113.

Rajashekar C, Burke MJ.1982. Liquid water during slow freezing based on cell water relations and limited experimental testing. In A Sakai, PH Li, eds, Plant Cold Hardiness and Freezing Stress. Academic Press, New York, pp 211-220.

Reitz, H.J. and W.T. Long. 1955. Water table fluctuation and depth of rooting of citrus trees in the Indian River area. *Proc. Fla. State Hort. Soc.* 68:24-29.

Rogers, J.S., and J.F. Bartholic. 1976. Estimated evapotranspiration and irrigation requirements for citrus. *Proc. Soil and Crop Sci. Soc. Fla.* 35:111-117.

Rogers, J.S., L.H. Allen, and D.V. Calvert. 1983. Evapotranspiration from a developing citrus grove in himid climate. *Trans. Am. Soc. Agric. Eng.* 26:1778-1783.

Sakai A, Larcher W (1987) Frost survival of plants: Responses and adaptation to freezing stress. Springer-Verlag, Berlin.

Slatery, R.O. 1967. Plant-water relationships. Academic Press, London and New York.

Smajstrla, A.G., L.R. Parsons, F.S. Zazueta, G. Vellidis, and K. Aribi. 1986. Water use and growth of young citrus trees. Paper No. 86-2069. Am Soc. Agric. Engr. Summer meeting, San Luis Obispo, CA.

Smith, P.F. 1956. Effect of phosphate fertilization on root growth, soil pH, and chemical constituents at different depths in an acid sandy Florida soil. *Proc. Fla. State Hort. Soc.* 69: 25-29.

Smith, P.F. 1965. Effect of nitrogen source and placement on the root development of 'Valencia' orange trees. *Proc. Fin. State Hort. Soc.* 78: 55-59.

Starr, J.L., and I.C. Paltineanu. 1998. Soil water dynamics using multisensor capacitance probes in nontraffic Interrow of Corn. *Soil Sci Soc. Am. J.* 62:114-122.

Stewart, E.H., J.E. Browning, and E.O Burt. 1969. Evapotranspiration as affected by plant density and water-table depth. *Trans. Am. Soc. Agric. Engr.* 12(5):646-647.

Sumner, D.M. 1996. Evapotranspiration from successional vegetation in a deforested area of the Lake Wales Ridge, Florida. *Water-Resources Investigation.* 96-4244.

Syvertsen, J. P. R.M. Zablotowicz, and M. L. Smith, JR. 1983. Soil Temperature and flooding on two species of citrus. *Plant and Sol* 72(3): 3-12.

Tindall, J. A., and J. R. Kunkel. 1999. Unsaturated Zone Hydrology for Scientists and Engineers. Prentice-Hall, Upper Saddle River, New Jersey.

Tinker, P. B., and P. H. Nye. 2000. Solution Movement in the Rhizosphere. Oxford University Press, New York, New York.

Tucker, D. P., C. G. Erickson, and K. T. Morgan. 1997. Middles Management Methods in Citrus Affect Soil Moisture Retention and Vegetation Species. *Proc. Fla. State Hort. Soc.* 110: 39-43.

Van den Honert, T. H. 1948. Water Transport in Plants as a Centenary Procrss. Discuss. *Faraday Soc.* 3: 146-153.

Wiegand, C.L., and W.A. Swanson. 1982a. Citrus responses to irrigation: I. Irrigation requirements; daily, monthly, and annual evapotranspiration amounts; and water management recommendations. J. *Rio Grande Valley Hort. Soc.* 35:73-85.

Wiegand, C.L., and W.A. Swanson. 1982b. Citrus responses to irrigation: II. Fruit yield, size, and number. *J. Rio Grande Valley Hort. Soc.* 35:87-95.

Wiegand, C.L., and W.A. Swanson. 1982c. Citrus responses to irrigation: III. Tree trunk and canopy growth. *J. Rio Grande Valley Hort. Soc.* 35:97-107.

Yelenosky, G., C. 1978. The effect of withholding water on cold hardiness of 'Valencia' orange and 'Star Ruby' grapefruit trees in controlled freezes. *Proc. Fla. State Hort. Soc.* 91: 18-20.

Yelenosky, G., C. 1985. Cold hardiness in citrus. *Hort. Rev*. 7: 201-238.

Yelenosky, G., C. 1996. An overview of Florida citrus freeze survival. *Proc. Fla. State Hort. Soc.* 109: 118-123.

Yelenosky, G., C., J. Hearn, and D. J. Hutchison. 1984. nonhardening temperatures- Major factor in freeze damage to citrus trees in December 1983. *Proc. Fal. State Hort. Soc*. 97: 33-36.

Young, R. and W.D. Bell. 1974. Photosynthesis in detached leaves of cold hardened citrus seedlings. *J. Amer. Soc. Hort. Sci*. 99: 400-403.

In: Drip Irrigation
Editor: Alfred H. Steele

ISBN: 978-1-63463-737-4

Chapter 2

MOLECULAR CHARACTERIZATION OF BIOFILMS DEVELOPING IN DRIP IRRIGATION SYSTEMS OPERATING WITH RECLAIMED WATER IN GRAN CANARIA (SPAIN)*

Olga Sánchez[1,*,†], Laura Garrido[1], Luis Fernando Arencibia Aguilar[2], Jordi Mas[1] and María del Pino Palacios-Díaz[3]

[1]Departament de Genètica i Microbiologia,
Universitat Autònoma de Barcelona, Bellaterra, Spain
[2]Servicio Técnico de Medio Ambiente y Emergencias,
Las Palmas de Gran Canaria, Spain
[3]Departamento de Patología Animal,
Producción Animal y Ciencia y Tecnología de los Alimentos,
Universidad de Las Palmas de Gran Canaria, Arucas, Spain

ABSTRACT

The traditional scarcity of hydric resources in Spain linked with a growing trend in water demand for applications such as landscaping, gardening or agricultural irrigation is gradually shifting water management practices towards the use of reclaimed water obtained from wastewater treatment plants. From a sustainable point of view, reclaimed water irrigation must be not only sanitary safe, but also should avoid problems in irrigation networks. Drip irrigation allow us to prevent sanitary risk, even when using cheaper and less intensive water treatments. But, on the other hand, this irrigation system is more sensitive to obstruction problems. Most of the problems are related to biofilm formation at the inner surfaces as well as to the points of delivery, and a crucial step for any antifouling strategy is to characterize the microbial communities involved in its development.

---

* These authors contributed equally to this work.

† Corresponding author. Email: olga.sanchez@uab.es

In this chapter we present the molecular characterization of the microbial assemblages in biofilms from drip irrigation systems in Gran Canaria (Spain) using treated wastewater from two wastewater treatment plants (WWTP), Bocabarranco and Cardones, presenting low and medium content of suspended solids (22.8 and 30.6 $mg \cdot L^{-1}$) respectively. Different samples (new drippers and older ones with more than 5 years of use) for irrigation of road green zones were examined from clogged and unclogged drippers. They were analyzed by denaturing gradient gel electrophoresis (DGGE) and subsequent sequencing of amplified 16S rRNA gene bands. DGGE patterns showed the presence of a large number of bands, indicating that the microbial community was rather complex. Most of the retrieved DGGE bands belonged to the phylogenetic groups Alpha and Gamma-proteobacteria, Bacteroidetes, Acidobacteria, Chlorobi, Nitrospirae and Firmicutes, but sequences from Beta and Delta-proteobacteria, Actinobacteria, Chloroflexi, Planctomycetes, Gemmatimonadetes, Deinococcus-Thermus and Cyanobacteria were also found.

Our results demonstrated that biofilms diversity was related with clogging phenomena, as higher diversity was found in the case of higher obstruction problems. Besides, there was an influence of dripper design on emitter obstruction (integral drippers were less prone to be clogged), a fact favoured by the presence of certain genera of bacteria. Besides, the results showed that the clogging process was able to occur in a short period of time in sensitive drippers.

In general, the biofilms formed in these irrigation systems using treated wastewater were more complex than we initially expected, with representation of groups usually found in aquatic environments. They showed a high diversity of unknown and uncultured microorganisms, suggesting that well structured biofilms resistant to pressure flushing could develop.

INTRODUCTION

The use of reclaimed water in the last decades, especially in those areas with low rainfall and a traditional water scarcity like the Mediterranean region, has become one of the best alternatives for agronomic purposes, such as landscaping, gardening or agriculture. However, if no proper water management is applied, reuse of effluents can lead to health risks associated with the presence of pathogen organisms such as bacteria, viruses, protozoa or helminths due to a possible human direct contact, the diffusion of aerosols (mainly in sprinkler irrigation), or to consumption of products irrigated with contaminated water. Drip irrigation, also called trickle irrigation, is particularly suitable for reclaimed water reuse since it minimizes the health risks to farmers and consumers because of the little contact of the product with water (Oron et al., 1996; Bahri, 1999). The physical and mechanical properties of soils, such as dispersion of particles, stability of aggregates, soil structure and permeability, are very sensitive to the type of exchangeable ions present in irrigation water. Thus, when effluent use is being planned, several factors related to soil properties must be taken into consideration (Ayers & Westcot, 1985), and in this sense, low contact in drip irrigation offers an advantage comparing with other irrigation systems. Other associated advantages are increased irrigation efficiency (there are practically non-existent water losses by runoff and deep percolation), minimization of the effect of salinity, improvement in the application of chemicals (fertigation and pest treatment) on crops, reduction of energy requirements and improved cultural practices (Ayars et al., 2007).

Nevertheless, the usage of reclaimed water for drip irrigation is often linked to clogging events that impair proper operation of the system, a phenomenon which causes a reduction in flow and consequently worsening irrigation uniformity, a process that has been extensively studied (Capra & Scicolone, 1998, 2007; Haijun & Guanhua, 2009; Puig-Bargués et al., 2010; Taylor et al., 1995; Zhang et.al., 2007). Measures of drip irrigation system performance include the determination of parameters as distribution uniformity (DU). Pitts (1997) established that only irrigation systems with DU higher than 87% can be considered as excellent ones. For this purpose, Uniformity Coefficient CU_{Chr} developed by Christiansen (1942), is experimentally measured. Basically, the main reasons causing obstruction of drippers can be classified into three groups: physical, chemical and biological (Bucks et al., 1979). Inorganic particles (sand, plastic, clay, etc) and organic materials (animal or microbiological waste) are the main causes of physical clogging, while chemical obstruction can be due to solid constitutents that interact forming precipitates. On the other hand, microbial activity is the origin of biological clogging, and it is related to biofilm formation (biofouling) (Gilbert et al., 1981; Taylor et al., 1995).

The effects of obstruction can be different depending on the dimensions of drippers (Ahmed et al., 2007) and the position thereof on the line (Ravina et al., 1997). On the other hand, Kreij et al. (2003) found that pipes working in laminar flux suffered more severe obstructions than those working in turbulent flux. The design of the emitters is also a key factor in their susceptibility for clogging.

Capra and Scicolone (2004) found that emitters with diffuser type (vortex) were more prone to clogging than those with labyrinth, and many authors agree that drippers have lower obstructions with high flow rates. Regarding water quality, Capra and Scicolone (2004) considered only two parameters: suspended solids and biochemical oxygen demand (BOD), to determine the risk of clogging when reclaimed water is used.

Therefore, as sustainable water management is needed, and given that when clogging occurs more than 90% of the accumulated material has a biological basis, the control of biological clogging is critical. As a consequence, a crucial step for any antifouling strategy is to characterize the microbial communities developing in this kind of facilities. Many studies have investigated the obstruction problems caused by biofilms during reutilization of wastewater for irrigation (Capra & Scicolone, 2004; Lubello et al., 2004; Sánchez et al., 2014; Tarchitzky et al., 2013; Yan et al., 2009), but only a few works attempted to describe the composition of the microbial assemblages present in these systems.

Among them, Yan et al. (2009) used scanning electron microscopy (SEM) in order to analyze the matrix structure of biofilms from drip irrigation emitters distributing reclaimed water, and Sánchez et al. (2014) concluded by means of molecular tools that potentially thermophilic microorganisms were involved in biofilms from greenhouse-enclosed drip irrigation systems. However, despite the importance this problem has in irrigation, little information exists concerning the microbial populations present in biofilms from drip irrigation installations.

In this work, we describe the microbial communities forming biofilms in the irrigation drippers used in road green zones in Gran Canaria (Canarian Islands, Spain) fed with reclaimed water from two different wastewater treatment plants (WWTP), Bocabarranco and Cardones. A previous field study showed that the facilities were prone to develop biofouling and clogging of the drippers due to lack of maintenance practices associated with a deficient irrigation system management. In that study, Arencibia et al. (2009) and Arencibia (2011) characterized the irrigation system operation depending on the dripper type (integrated in drip line, "integral", vs on line drippers), its time of operation (new ones vs older than 5 years), and the reclaimed water origin (from both of aforementioned wastewater treatment plants), trying to recommend the use of better management practices to avoid the problems caused by clogging.

Therefore, they detected irrigation lines with clogged drippers and unclogged ones. In the present study, and in order to gain a better understanding of biofouling, both clogged and unclogged drippers were sampled after cleaning the system, and were examined by means of denaturing gradient gel electrophoresis (DGGE) (Muyzer et al., 1998), a fingerprinting technique useful to obtain patterns of natural microbial communities and to characterize the more abundant members of these populations. The information obtained will contribute to improve management strategies to minimize the problems associated to biofouling in microirrigation, especially when using reclaimed water.

MATERIALS AND METHODS

Description of the Irrigation System

The irrigation system was used in road green zones from the Gran Canaria Island (Spain), and it was fed with reclaimed water from two WWTPs, Bocabarranco and Cardones, which treat 2,000 and 3,000 $m^3 \cdot day^{-1}$ of urban wastewater respectively.

An extensive description of the studied system was presented by Arencibia (2011), who characterized the standard chemical parameters of these reclaimed waters (Table 1).

Table 1. Suspended solids (SS) and biochemical oxygen demand (BOD) of the reclaimed water used from Cardones and Bocabarranco WWTPs, in $mg \cdot L^{-1}$ (Std Dev: standard deviation, CV: coefficient of variation)

WWTP	SS ($mg \cdot L^{-1}$)			BOD ($mg \cdot L^{-1}$)		
	Mean	Std Dev	CV	Mean	Std Dev	CV
Cardones	30,60	5,18	16,92	26,20	16,18	61,74
Bocabarranco	22,80	11,03	48,38	18,40	7,86	42,72

Evaluation of Emitter Clogging

For each experimental unit, the outflow (Q) was gauged at 4 points (start, 1/3, 2/3 and end of line) in 4 lines during five minutes, measuring the amount of water collected per dripper, all of them having a nominal flow rate of 4 $L \cdot h^{-1}$. Then, Christiansen Uniformity Coefficient, CU_{Chr}, (Christiansen, 1942) was calculated as:

$$CU_{Chr} = \frac{Q_{(25\%)}}{Q_{mean}} * 100$$

where $Q_{(25\%)}$ is the water flow of the 25% lower flow drippers, and Q_{mean} is the total mean flow.

Also, pressures at the beginning and the end of each line were measured using a "Bourdon" glycerin gauge (accuracy of 0.2 bar) and the coefficient of variation (CVp) was calculated.

Sampling

Clogged (O) and unclogged (B) drippers from the studied lines were localized. The following drippers were sampled after flushing the irrigation system: new on line emitters, "Azud-8", (less than one year of functioning, samples 1 and 2) vs older than 5 years (samples 3 and 4), as well as one integrated emitter "Azudrip" (sample 5) from Bocabarranco WWTP. From Cardones WWTP, two new dripper types were sampled: on line drippers (samples 6 and 7) vs emitters integrated in drip lines (samples 8 and 9) (Table 2).

Biofilms from the plastic surfaces of the drippers were scraped and collected under sterile conditions using a sterilized metal spatula and stored in an eppendorf tube at -20ºC until further analysis.

Table 2. Clasification of samples depending on the water origin, the dripper type and the dripper age

Sample code	WWTP	Dripper type	Dripper age
3O	Bocabarranco	on line	old
3B	Bocabarranco	on line	old
4O	Bocabarranco	on line	old
1O	Bocabarranco	on line	new
1B	Bocabarranco	on line	new
2O	Bocabarranco	on line	new
2B	Bocabarranco	on line	new
5O	Bocabarranco	integral	new
6O	Cardones	on line	new
6B	Cardones	on line	new
7O	Cardones	on line	new
8O	Cardones	integral	new
8B	Cardones	integral	new
9O	Cardones	integral	new
9B	Cardones	integral	new

DNA Extraction

The DNA Power Soil Kit MO BIO ref. 1288-50 (MO BIO Laboratories, Inc., USA) was used for DNA extraction from biofilms. DNA integrity was checked by agarose gel electrophoresis and quantified using a low DNA mass ladder as a standard (Invitrogen, Life Technologies, USA) using the Quantity One software package (Bio-Rad, Spain) for gel documentation and analysis.

PCR-DGGE Fingerprinting

Fragments of the bacterial 16S rRNA gene suitable for DGGE analysis were obtained by using the specific primer 358F with a 40-bp GC clamp, and the universal primer 907RM as described in Sánchez et al. (2007). Polymerase chain reaction (PCR) was carried out with a Biometra thermocycler using the following program: initial denaturation at 94ºC for 5 min; 10 touchdown cycles of denaturation (at 94ºC for 1 min), annealing (between 63-53 ºC for 1 min, decreasing 1ºC each cycle), and extension (at 72ºC for 3 min); 20 standard cycles (annealing at 55.5ºC, 1 min) and a final extension at 72ºC for 5 min. PCR mixtures contained the template DNA, each deoxynucleoside triphosphate at a concentration of 200 μM, 1.5 mM $MgCl_2$, each primer at a concentration of 0.3 μM, 2.5 U Taq DNA polymerase (Invitrogen) and the PCR buffer supplied by the manufacturer. BSA (Bovine Serum Albumin) was added at a final concentration of 600 $\mu g \cdot ml^{-1}$ to minimize the inhibitory effect of humic substances (Kreader, 1996). The volume of reactions was 50 μl. PCR products were verified and quantified by agarose gel electrophoresis with a low DNA mass ladder standard (Invitrogen) using the Quantity One software package (Bio-Rad).

DGGEs were run in a DCode system (Bio-Rad, Spain) as described by Muyzer et al. (1998) using a 6% polyacrylamide gel with a gradient of 30-70% DNA-denaturant agent. Seven hundred ng of PCR product were loaded for each sample and the gels were run at 100 V for 18 h at 60°C in 1xTAE buffer (40 mM Tris [pH 7.4], 20 mM sodium acetate, 1 mM EDTA). The gel was stained with SybrGold (Molecular Probes, Life Technologies, USA) for 45 min, rinsed with 1xTAE buffer, removed from the glass plate, placed in a UV-transparent gel scoop, and visualized with UV in a Gel Doc XRS (Bio-Rad, Spain).

Analysis of DGGE Patterns and Statistical Analyses

Digitized DGGE images were analyzed using Quantity One from Bio-Rad (Spain). Bands occupying the same position in the different lanes of the gel were identified. A matrix was constructed for all lanes, taking into account the presence or absence of the individual bands. This matrix was used to obtain a dendrogram comparing samples by means of the unweighted-pair group method using the Dice coefficient with the software SPSS.

In order to obtain direct descriptors of the diversity of bacterial assemblages, we calculated the Shannon index (*H'*), as explained by Magurran (1988). It was calculated using the following function:

$$H' = -\sum_{i=1}^{i=n} p_i \ln p_i$$

where n is the number of species in a sample (number of DGGE bands in our study) and p_i the proportion of a certain species (relative band intensity).

16S rRNA Gene Sequencing and Analyses

Visible bands were excised from the gels, resuspended in milli-Q water overnight and reamplified for sequencing. Purification of PCR products from DGGE bands and sequencing reactions was performed by Macrogen (South Korea) with primer 907RM. They utilized the Big Dye Terminator version 3.1 sequencing kit and reactions were run in an automatic ABI 3730XL Analyzer-96 capillary type. Sequences were subjected to a BLAST search (Altschul et al., 1997) to obtain an indication of the phylogenetic affiliation Gene sequences were deposited in GenBank under accession numbers LM993957-LM994028.

RESULTS AND DISCUSSION

Evaluation of Emitter Clogging

Christiansen Uniformity Coefficients (CU_{Chr}) and pressure (P) of the lines in which the drippers were sampled are presented in Table 3. Also, the rate between water flow from clogged and unclogged drippers of each line before (B) and after (A) flushing was calculated, and expressed in %. As observed, the worst values of CU_{Chr} (60.5 %) were obtained in the old on line drippers of Bocabarranco operated with a pressure of 1.7 bar. On the other hand, low CU_{Chr} values (65.6 %) of integral drippers of Cardones are due to the low pressure (0.5 bar) in which the irrigation system was operated (far from the nominal pressure of emitters) and not due to the existence of important clogging problems.

Table 3. Christiansen Uniformity Coefficient (CU_{Chr}) and mean Pressure (P) of the lines in which drippers were sampled. The coefficient of variation CV_p was also calculated. Rates between water flow from clogged and unclogged drippers of each line (O/B) before (B) and after (A) flushing are also presented. Boc: Bocabarranco WWTP, C: Cardones WWTP

WWTP	Dripper type	Dripper age	CU_{Chr}	meanP	CV_P	samples	O/B(%)B	O/B(%)A
Boc	On line	old	60.5	1.7	4.5	3	24	na
						4	75	na
Boc	On line	new	85.6	2.2	12.7	1	7	32
						2	68	na
Boc	Integral	new	90.8	0.7	29.8	5	na	na
C	On line	new	96.1	0.7	3.5	6	81	82
						7	80	86
C	Integral	new	65.6	0.5	33.9	8	85	103
						9	93	100

na: not available.

Microbial Community Analyses

Samples from different biofilms developing in drip irrigation systems in Gran Canaria, fed with treated wastewater from two WWTPs (Bocabarranco and Cardones), were analyzed by Denaturing Gradient Gel Electrophoresis (DGGE), a molecular tool which allows a reasonably straightforward comparison of the phylogenetic composition of a large number of samples along space and time (Figure 1). This technique results in a fingerprint of bands, each band corresponding to an operational taxonomic unit (OTU). Besides, DGGE further allows the assessment of the diversity of the assemblage by subsequent sequencing.

Band number per lane ranged between 24 and 52 (mean 39) in Bocabarranco biofilms, and between 17 and 39 (mean 30) in Cardones biofilms (Figure 1), suggesting a higher diversity in the case of biofilms fed with Bocabarranco WWTP, which presented higher clogging problems (Table 3). In general, DGGE patterns showed the presence of a large number of bands, indicating that the microbial communities were rather complex.

Bands were excised from the gels and sequenced to determine their phylogenetic affiliation (Table 4). Informative sequences were obtained from 34 bands in biofilm samples from Bocabarranco and 38 bands in biofilm samples from Cardones. These bands accounted for 65.4% (Bocabarranco) and 74.2% (Cardones) of the total mean band intensity, indicating that most of the bacterial diversity could be identified. In general, most of the sequences were closely related to 16S rRNA gene sequences from uncultured microorganisms, although we could determine which major phylogenetic group they belonged to, while others matched well with cultured bacteria. The similarities ranged between 81.1 and 100%.

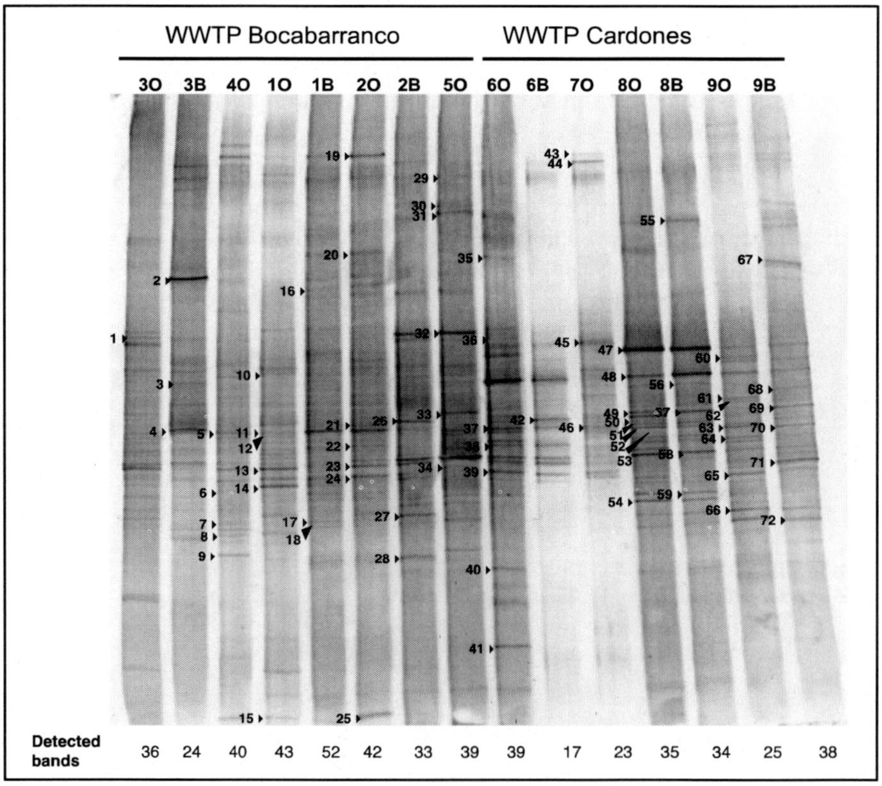

Figure 1. DGGE image showing the fingerprints of microbial communities forming biofilms in the irrigation drippers used in road green zones in Gran Canaria (Canarian Islands, Spain) fed with treated wastewater from two different WWTPs, Bocabarranco and Cardones. Numbers indicate different drippers, as listed in Table 2. The number of detected bands per lane are also indicated. O samples: clogged drippers, B samples: unclogged drippers.

Table 4. Phylogenetic identification of sequences obtained from dripper biofilm DGGE bands

Band	Closest match	% similarity (n°bases)[a]	Taxonomic group	Accession n° (Gen Bank)	Cultured closest match (% similarity)
1	Clone AnDHS-A10	88.5 (430)	Chlorobi	AB359015	Chlorobi Bacterium Mat9-16 (86.0)
2	Clone B15_130	97 (521)	γ-proteobacteria	EU790281	*Legionella*-like Amoebal Pathogen (89.1)
3	*Ochrobactrum tritici* strain S107	93.5 (386)	α-proteobacteria	AY972352	
4	*Pseudoxanthomonas mexicana*	95.9 (473)	γ-proteobacteria	GQ480839	
5	Clone H2SRC24x	89.9 (472)	γ-proteobacteria	FM213074	*Dokdonella* sp. PYM5-8 (88.6)
6	Uncultured *Nocardioles* sp. Clone AUVE_05B01	95 (497)	Actinobacteria	EF651193	*Marmoricola* sp. MK05 (94.6)
7	Clone Lehavim-241	93 (508)	Actinobacteria	GC425748	*Solirubrobacter* sp. L64 (92.1)
8	Uncultured Actinobacterium Clone H5Ba05	94 (505)	Actinobacteria	AY360557	*Solirubrobacter* sp. L64 (90.4)
9	Clone L/142-3p23	98.5 (471)	Chloroflexi	FJ72381	*Sphaerobacter termophilus* (95.0)
10	Clone HH2_h4	85.4 (449)	Acidobacteria	FN401235	Bacterium ellin 6075 (85.0)
11	Uncultured acidobacteriaceae Clone AMFG6	81.3 (404)	Acidobacteria	AM935271	Acidobacteria bacterium (79.3)
12	Clone Sbrh_192	88 (461)	β-proteobacteria	FJ175147	*Ideonella* sp. B513 (87.8)
13	Uncultured *Nitrospira* sp. Clone as2-4	99.1 (524)	Nitrospirae	GU257604	Candidatus *Nitrospira defluvii* (99.1)
14	Uncultured *Nitrospira* sp. Clone GASP-WB2W3C05	99.6 (529)	Nitrospirae	EF074297	*Nitrospira* sp. (99.6)
15	Uncultured Sphaerobacteriaceae clone AMAH5	96.0 (506)	Chloroflexi	AM935838	*Sphaerobacter termophilus* (88.7)

Table 4. (Continued)

Band	Closest match	% similarity (n°bases)[a]	Taxonomic group	Accession n° (Gen Bank)	Cultured closest match (% similarity)
16	Uncultured Cyanobacterium isolate DGGE band 6	99.2 (516)	Stramenopiles	AY942895	*Nannochloropsis granulata* (95.0)
17	*Thiocapsa* sp.	92.0 (487)	γ-proteobacteria	FN293074	
18	Clone A6_25	86.6 (460)	γ-proteobacteria	AM940543	*Hydrocarbonipha ga efusa* (76.4)
19	Uncultured Cyanobacterium clone SW-xj279	98.6 (515)	Stramenopiles	GQ302545	*Nanofrustulum shiloli* (98.3)
20	Uncultured Bacteroidetes bacterium clone 4P23	99.1 (530)	Bacteroidetes	AJ871049	*Chitinophaga* sp. (94.4)
21	Clone nbw42g10cl	90.4 (492)	β-proteobacteria	GQ060002	*Ideonella* sp. (90.1)
22	Uncultured *Lysobacter* sp. Clone Plot4-A02	88.6 (467)	γ-proteobacteria	EU449639	*Lysobacter yangpyeongensi* (88.4)
23	Candidatus *Nitrospira defluvii*	100 (530)	Nitrospirae	GU454943	
24	Uncultured *Nitrospira* sp. Clone Vm1	99.6 (528)	Nitrospirae	GU229885	*Nitrospira* cf. *moscoviensis* (99.4)
25	Clone OS-21	97.1 (507)	Chloroflexi	AB205952	*Roseiflexus* sp. (87.0)
26	Clone 13C-DNA-AOB-clone 17	99.8 (540)	Nitrospirae	AB475005	*Nitrospira* sp. (99.8)
27	Clone G13	92.7 (486)	δ-proteobacteria	AF407700	*Geothermobacter* sp. (86.1)
28	*Meiothermus* sp.	100 (508)	Deinococcus-Thermus	EU149044	
29	Clone FW1022-033	96.7 (505)	Chlorobi	EF693117	Chlorobi Bacterium Mat9-16 (88.3)
30	Clone K29C2-25	96.7 (492)	Chlorobi	AB504656	Chlorobi Bacterium Mat9-16 (88.3)
31	Uncultured Chlorobi bacterium	99.6 (527)	Chlorobi	CU926515	Chlorobi Bacterium Mat9-16 (88.3)
32	Chlorobi bacterium Mat9-16	97.2 (522)	Chlorobi	AB478415	

Band	Closest match	% similarity (n°bases)[a]	Taxonomic group	Accession n° (Gen Bank)	Cultured closest match (% similarity)
33	Clone F24_Pitesti	94.6 (492)	γ-proteobacteria	DQ378188	*Thiorhodospira* sp. (87.3)
34	Clone OT90-13_org	93.5 (474)	Deinococcus-Thermus	FJ945747	*Meiothermus ruber* (92.7)
35	Uncultured Sphingobacteriales clone GASP-KC3W2_F09	95.1 (500)	Bacteroidetes	EU300519	*Flavosolibacter* sp. (93.6)
36	Clone FFCH16449	83.3 (438)	Bacteroidetes	EU133728	*Sphingobacterium* sp. (82.7)
37	Clone SX2-1	90.0 (466)	β-proteobacteria	DQ469206	*Georgfuchsia toluolica* (87.7)
38	*Thermomonas dokdonensis*	90.6 (475)	γ-proteobacteria	EF100698	
39	Clone AO6day03C-ARISA791	95.6 (481)	Nitrospirae	GU933963	Candidatus *Nitrospira defluvii* (95.2)
40	Clone nbw863a11c1	89.0 (459)	Gemmatimonadetes	GQ029564	Bacterium Ellin5220 (86.8)
41	Clone 10D-4	89.4 (463)	Chloroflexi	DQ906857	*Kouleothrix aurantiaca* (88.9)
42	Uncultured Rhodocyclaceae clone REG_R2P2_G1	89.2 (473)	β-proteobacteria	FJ933281	*Kouleothrix aurantiaca* (88.4)
43	*Elizabethkingia miricola*	98.9 (518)	Bacteroidetes	GQ398109	
44	Clone MFC-EB32	99.6 (517)	Bacteroidetes	AJ630300	Sphingobacteriaceae bacterium (94.5)
45	*Acinetobacter* sp.	100 (533)	γ-proteobacteria	HM047743	
46	Uncultured gamma proteobacterium clone GASP-WA153_C11	96.6 (511)	γ-proteobacteria	EF072205	*Frateuria* sp. (95.5)
47	Clone XJ118	100 (528)	Chlorobi	EF648160	Chlorobi bacterium Mat9-16 (99.4)
48	Clone Bas-7-67	96.8 (488)	Acidobacteria	GQ495424	Bacterium Ellin6075 (96.8)
49	Uncultured *Burkholderia* sp. Clone T8116	99.1 (527)	β-proteobacteria	EU029570	*Aquabacterium* sp. (97.2)
50	Clone NO17ant18f06	97.4 (517)	β-proteobacteria	GQ921462	*Georgfuchsia toluolica* (93.8)

Table 4. (Continued)

Band	Closest match	% similarity (n°bases)[a]	Taxonomic group	Accession n° (Gen Bank)	Cultured closest match (% similarity)
51	Uncultured Rhodocyclaceae bacterium clone 408	95.0 (508)	β-proteobacteria	FM207908	*Georgfuchsia toluolica* (92.7)
52	Clone ACS29	89.3 (469)	β-proteobacteria	FJ375442	*Caldimonas manganoxidans* (88.8)
53	*Thermomonas haemolytica*	98.7 (516)	γ-proteobacteria	NR_025441	
54	Uncultured *Nitrospira* sp. Clone REV_R1PII_8B	100 (520)	Nitrospirae	FJ933472	*Nitrospira* sp. (92.7)
55	Clone AOA_129	100 (519)	Bacteroidetes	AB479737	*Flexibacter* sp. (87.3)
56	Uncultured Acidobacteria	93.6 (470)	Acidobacteria	AM902634	Bacterium Ellin6099 (89.7)
57	*Exiguobacterium* sp.	99.4 (525)	Firmicutes	HM016870	
58	Clone KIST-JJY016	84.8 (441)	Nitrospirae	EF594049	*Nitrospira* sp. (84.7)
59	Uncultured Planctomycete clone 5GA_Pla_HKP_31	98.0 (496)	Planctomycetes	GQ356177	
60	Clone MA00070F02	98.4 (493)	Acidobacteria	FJ772353	Bacterium Ellin6075 (97.8)
61	*Bradyrhizobium* sp.	98.8 (492)	α-proteobacteria	AB367691	
62	*Bradyrhizobium* sp. ALG3	81.1 (408)	α-proteobacteria	EU871483	
63	Clone Kab133	100 (528)	γ-proteobacteria	FJ936850	*Frateuria* sp. (99.8)
64	*Sphingomonas* sp.	94.0 (471)	α-proteobacteria	FN794222	
65	Clone FFCH6108	95.3 (489)	Acidobacteria	EU132432	Bacterium Ellin6075 (86.4)
66	Uncultured Actinobacterium clone H5Ba05	94.8 (498)	Actinobacteria	AY360557	Bacterium Ellin301 (90.8)
67	Clone SS-134	97.9 (511)	Bacteroidetes	AY945889	*Flavisolilbacter ginsengisoli* (94.5)
68	Clone ncd1067b08c1	99.6 (527)	Firmicutes	HM335937	*Bacillus megaterium* (99.6)
69	*Exiguobacterium* sp.	100 (529)	Firmicutes	HM343561	
70	Clone LaYa6-62	99.0 (520)	Firmicutes	GU291682	*Bacillus* sp. (99.0)
71	*Nitrospira* sp.	99.2 (510)	Nitrospirae	AJ224038	

Band	Closest match	% similarity (nºbases)[a]	Taxonomic group	Accession nº (Gen Bank)	Cultured closest match (% similarity)
72	Clone ncd761d08c1	97.9 (518)	Actinobacteria	HM299603	*Solirubrobacter* sp. (87.4)

Taking into account the relative intensity of each group in the different lanes, their contribution (%) could be calculated and summarized in Figure 2 for each lane. A percentage of bands could not be identified due to either a failure to reamplify faint bands or the poor quality of some sequences. Some phylogenetic groups, like Gamma-proteobacteria, Acidobacteria and Nitrospirae could be retrieved in all the samples, while others (Alpha-proteobacteria, Bacteroidetes, Chlorobi and Firmicutes) were present in most of them. Sequences from Beta and Delta-proteobacteria, Actinobacteria, Chloroflexi, Planctomycetes, Gemmatimonadetes, Deinococcus-Thermus and Cyanobacteria were also found, although in a minor number of samples. The results showed that the biofilms had a heterogeneous composition with representation of groups usually found in aquatic environments. Some of the retrieved bands had a high similarity with genus involved in the nitrogen cycle, like *Nitrospira* sp. (Nitrospirae), which is a nitrite-oxidizing bacterium often found in wastewater treatment systems (Hovanec et al., 1998), or *Bradyrhizobium* sp. (Alpha-proteobacteria), a genus of Gram-negative soil bacteria with many species able to fix nitrogen. On the other hand, *Pseudoxanthomonas mexicana*, a Gamma-proteobacterium (Thierry et al., 2004), can reduce nitrite to N_2O. Other bands of cultured bacteria belonged to microorganisms involved in biofilm formation, like *Acinetobacter* sp. or *Sphingomonas* sp., widely distributed in nature. In drinking water, the genus *Acinetobacter* has been shown to contribute to bacterial aggregation, while *Sphingomonas* is a genus recognized by its capability to degrade a wide variety of refractory environmental pollutants and carry out diverse other biotechnologically useful activities, such as the biosynthesis of valuable polimers (Laskin & White, 1999). *Sphingomonas* sp. could also be involved in biofilm formation due to its ability to form exopolimers associated with the initial adhesion of bacteria, which is the primary step for biofilm formation (Azeredo & Oliveira, 2000).

One of the groups retrieved in all samples, with sequences from uncultured microorganisms, is Acidobacteria, a newly devised phylum, whose members are physiologically diverse and ubiquitous, especially in soils, but are under-represented in culture. Members of this phylum are physiologically diverse, some being acidophilic, and were first recognized as a novel division in 1997 (Kuske et al., 1997). Since they have only recently been discovered and the large majority have not been cultured, the ecology and metabolism of these bacteria is not well understood (Quaiser et al., 2003). However, these bacteria may be important contributors to ecosystems, since they are particularly abundant within soils.

Other uncultured sequences corresponded to the phylum Bacteroidetes, which is composed of three large classes of Gram-negative, nonsporeforming, anaerobic, and rod-shaped bacteria that are widely distributed in the environment, including soil, sediments, and seawater, as well as in the guts and on the skin of animals. Bacteroidetes are commonly assumed to be specialized in degrading high molecular weight (HMW) compounds and to have a preference for growth attached to particles and surfaces (Fernández-Gomez et al., 2013). One of the retrieved sequences of Bacteroidetes, however, had a high similarity to a cultured microorganism, *Elizabethkingia miricola*, an emergent opportunistic human pathogen associated with sepsis, which has also been isolated from diverse ecological niches, including eutrophic lakes, soil, freshwater sources, spent nuclear fuel pools, and water condensation spaces; recently, the ability of *Elizabethkingia* spp. to form biofilms in biotic or abiotic surfaces has been recognized (Jacobs & Chenia, 2011).

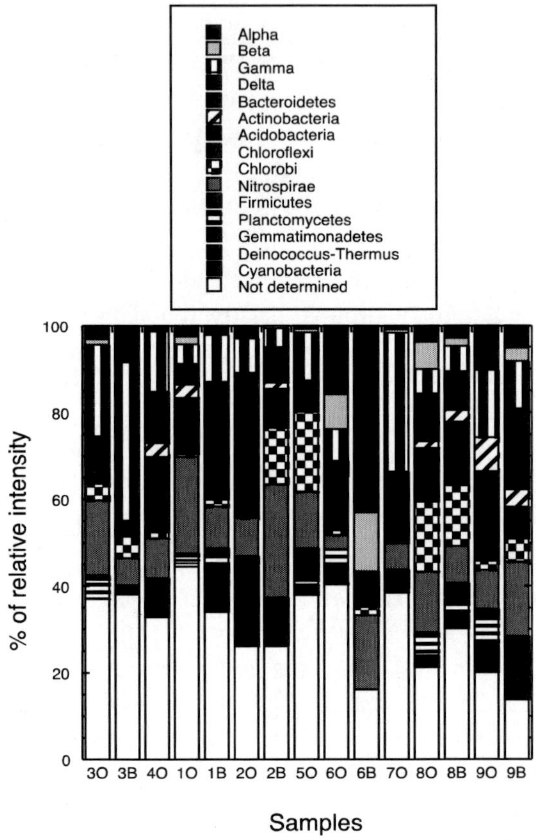

Figure 2. Percentage of relative intensity of DGGE bands affiliated to different phylogenetic groups. Numbers indicate different drippers, as listed in Table 2. O samples: clogged drippers, B samples: unclogged drippers.

Another phylum with retrieved sequences similar to cultured bacteria was Firmicutes, which are Gram-positive with a rigid cell wall structure (Desvaux et al., 2006). Many Firmicutes, like *Bacillus* spp., produce endospores, which are resistant to desiccation and can survive extreme conditions forming biofilms in drip irrigation systems (Sánchez et al., 2014).

On the other hand, Actinobacteria, although with sequences retrieved only in some samples, deserves some consideration. They are a distinct group of bacteria that are widely distributed in nature (Zucchi et al., 2011). Currently, Actinobacteria comprise eight groups with 48 genera (Holt et al., 1994). Special attention has been given to this bacterial group in biotechnological applications, which are a natural result of their great metabolic diversity (Piret & Demain, 1988). Actinobacteria are the most common source of antibiotics (Okami & Hotta, 1988) and are a promising source of a wide range of enzymes, enzyme inhibitors, immunomodifiers, and vitamins (Peczynska-Czoch & Mordarski, 1988). In nature,

Actinobacteria play an important role in the cycling of organic compounds and have been associated with soil organic matter production, including production of the black pigments called melanins, which are related to soil humic acid (Gomes et al., 1996).

A binary matrix was constructed for all lanes and used to obtain a dendrogram (Figure 3) based on UPGMA clustering (unweighted-pair group method using average linkages, Dice similarity coefficient). Concerning the effect of clogging in the microbial diversity patterns, the results indicate that in the case of the Cardones WWTP, with similar water flow between clogged and unclogged drippers (O/B >80%, Table 3), the pattern was not significantly different between O and B samples, with band patterns virtually identical (with the exception of sample 6O). Besides, Cardones samples clustered separately depending on the dripper type, with samples from lines with integral new drippers (8O, 8B, 9O and 9B) forming a group apart from samples from on line new drippers (6O, 6B and 7O), suggesting an influence of dripper design in the composition of the microbial assemblage. Notice that after flushing, OB_A in integral drippers (8 and 9) reached 100 %, which suggests that drippers were virtually clean, and had the same genera of bacteria.

In contrast, band patterns from the Bocabarranco WWTP, with different water flow between clogged and unclogged drippers (O/B <80%, Table 3) were significantly different depending on the state of the dripper sampled (clogged or unclogged), while the dripper age (old –more than 5 years of use- and new) did not seem to influence the microbial community structure. Notice that the worst values of O/B were obtained for samples 1 and 3 (7 and 24% respectively before flushing), having a very close pattern, and different from the rest of the samples. This fact suggest that those microorganisms prone to develop biofouling were growing in those samples. Therefore, It would be important to focus in the determination of all the species growing in samples 1O and 3O. In general, cluster analysis is coherent to describe the clogging phenomena. Also, cluster analysis point to the lack of consistent grouping around the WWTP variable. This probably indicates that the characteristics of the incoming water (with similar origen and water quality parameters) were not the main determinants of the microbial patterns observed, being probably the type of dripper a more crucial factor. Also, the clogging process seemed to occur in a short period of time in on line drippers, because their age was not a determining factor.

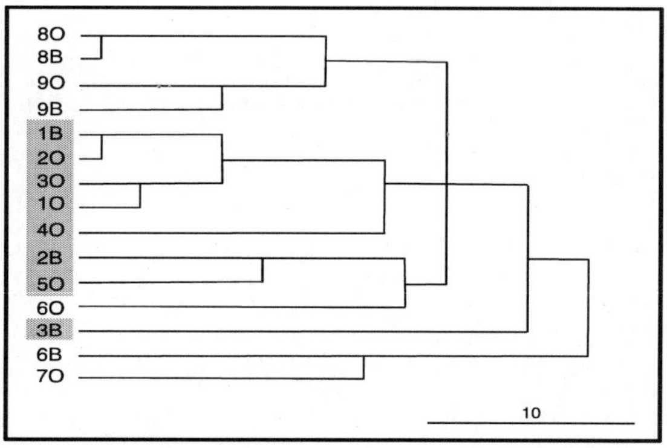

Figure 3. Euclidean-distance dendrograms generated from the DGGE profiles of the samples analyzed, determinated by the unweighted-pair group method using average linkages. The scale bar is linkage distance. Biofilm samples from drippers fed with water from Bocabarranco WWTP are highlighted with a gray square; the remaining samples correspond to biofilms fed with water from Cardones WWTP.

Diversity Index

For the sake of comparison, the Shannon diversity index was calculated from the DGGE fingerprints in each sample (Figure 4).

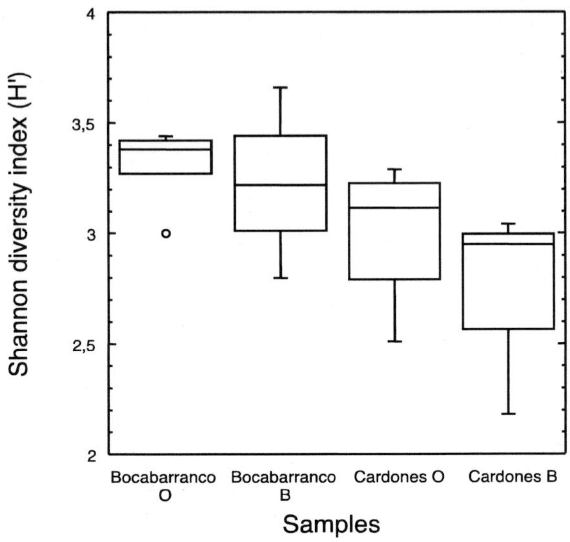

Figure 4. Boxplots of Shannon index for the samples from biofilms analyzed from clogged (O samples) and unclogged (B samples). Drippers were fed with treated wastewater from two different WWTPs, Bocabarranco and Cardones.

Shannon diversity indices, which provide information about the richness and the relative abundances of different species, ranged between 3.0 and 3.4 in Bocabarranco O samples, 2.9 and 3.7 in Bocabarranco B, 2.5 and 3.3 in Cardones O, and 2.1 and 3.0 in Cardones B. Many studies have measured this index in different systems (Adrados et al., 2014; Garrido et al., 2014; Liu et al., 2012; Lymperopoulou et al., 2012; Schauer et al., 2000; Vaz-Moreira et al., 2011) and values tipically range between 0.5 and 5, indicating low diversity with values from 0.5 to 2 and high diversity with values from 2 to 5. Our results showed intermediate values of diversity, being slightly higher in Bocabarranco biofilms, which presented higher clogging problems. This fact suggest that the emitter obstruction phenomena is favoured by the presence of certain genera of bacteria.

CONCLUSION

Our results demonstrated that biofilms diversity was related with clogging events, as higher diversity was found when important obstruction problems occurred. Also, there is an influence of dripper design in the composition of the microbial assemblages, being integral drippers less prone to be clogged. These facts suggest that emitter obstruction phenomena are favoured by the presence of certain genera of bacteria. Furthermore, the clogging process is able to occur in a short period of time in sensitive drippers.

In general, the microbial communities developed in drippers were rather complex, and the biofilms studied had a heterogeneous composition with representation of groups usually found in aquatic environments. The group Acidobacteria, whose members are physiologically diverse and ubiquitous, especially in soils, was retrieved in all samples, as well as the class Gammaproteobacteria. Genera involved in the nitrogen cycle, as well as microorganisms involved in biofilm formation were also identified. *Sphingomonas*, a genus recognized by its capability to degrade a wide variety of refractory environmental pollutants but also able to be involved in biofilm formation, Firmicutes, which can survive extreme conditions forming biofilms in drip irrigation systems, and Actinobacteria, which play an important role in the cycling of organic compounds, were also detected. Therefore, further attempts to isolate the key microorganisms involved in dripper clogging will be essential in order to explore future antifouling strategies.

ACKNOWLEDGMENT

This work was supported by the spanish project Consolider TRAGUA (CSD-2006-00044).

REFERENCES

Adrados, B., Sánchez, O., Arias, C. A., Bécares, E., Garrido, L., Mas, J., Brix, H. & Morató, J. (2014). Microbial communities from different types of natural wastewater treatment systems: Vertical and horizontal flow constructed wetlands and biofilters. *Water Res.*, 55, 304-312.

Ahmed, B. A. O., Yamamoto, T., Fujiyama, H. & Miyamoto, K. (2007). Assessment of emitter discharge in microirrigation system as affected by polluted wáter. *Irrig. Drain. Syst.*, 21, 97-107.

Altschul, S. F., Madden, T. L., Schäffer, A. A., Zhang, J.¡, Zhang, Z., Miller W., et al. (1997). Gapped BLAST and PSI_BLAST, a new generation of protein database search programs. *Nucleic Acids Res.*, 25, 3389-3342.

Arencibia, L. F., Fernández, F., Morant, T., Mendoza-Grimón, V. & Palacios-Diaz, M. P. (2009). Congreso nacional AERYD. Mantenimiento de sistemas de riego localizado con aguas regeneradas en zonas verdes de las carreteras de la isla de Gran Canaria.

Arencibia, L. F. (2011). Mantenimiento de sistemas de riego localizado con aguas regeneradas en zonas verdes de las carreteras de la isla de Gran Canaria. Master Thesis. Doctorado en Ingeniería Ambiental y Desalinización. Universidad de Las Palmas de Gran Canaria.

Ayars, J. E., Bucks, D. A., Lamm, F. R. & Nakayama, F. S. (2007). Chapter 1: Introduction. In F. R. Lamm, J. E. Ayars & F. S. Nakayama (Eds.), Microirrigation for crop production: design, operation and management (pp. 1-26). Amsterdam: Elsevier.

Ayers, R. S. & Westcot, D. W. (1985). Irrigation and Drainage. Paper Nº 29, Rev. 1 FAO.

Azeredo, J. & Oliveira, R. (2000). The role of exopolymers produced by *Sphingomonas paucimobilis* in biofilm formation and composition. *Biofouling,* 16, 17-27.

Bahri, A. (1999). Agricultural reuse of wastewater and global water management. *Water Sci. Technol.*, 4-5, 339-346.

Bucks, D. A., Nakayama, F. S. & Gilbert, R. G. (1979). Trickle irrigation water quality and preventive maintenance. *Agric. Water Manag.*, 2, 149-162.

Capra, A. & Scicolone, B. (1998). Water quality and distribution uniformity in drip/trickle irrigation systems. *J. Agric. Eng. Res.,* 70,355-365.

Capra, A. & Scicolone, B. (2004). Emitter and filter tests for wastewater reuse by drip irrigation. *Agric. Water Manag.*, 68, 135-149.

Capra, A. & Scicolone, B. (2007). Recycling of poor quality urban wastewater by drip irrigation systems. *J. Clean. Prod.er.*, 15, 1529-1534.

Christiansen, J. E. (1942). Irrigation by sprinkling. *California Agriculture Experiment Station Bulletin,* 670.

Desvaux, M., Dumas, E., Chafsey, I. & Hébraud, M. (2006). Protein cell surface display in Gram-positive bacteria: from single protein to macromolecular protein structure. *FEMS Microbiol. Lett.*, 256, 1-15.

Fernández-Gómez, B., Richter, M., Schüler, M., Pinhassi, J., Acinas, S. G., González, J. M., Pedrós-Alió, C. (2013). Ecology of marine Bacteroidetes: a comparative genomics approach. *ISME J.,* 7: 1026-1037.

Garrido, L., Sánchez, O., Ferrera, I., Tomàs, N. & Mas, J. (2014). Dynamics of microbial diversity profiles in waters of different qualities. Aproximation to an ecological quality indicator. *Sci. Total Environ.*, 468-469, 1154-1161.

Gilbert, R. G., Nakayama, F. S., Bucks, D. A., French, O. F. & Adamson, K. C. (1981). Trickle irrigation, emitter clogging and flow problems. *Agric. Water Manag.*, 3, 159-178.

Gomes, R. C. Mangrich, A. S., Coelho, R. R. R. & Linhares, L. F. (1996). Elemental, functional group and infrared spectroscopic analysis of actinomycete melanins from Brazilian soils. *Biol. Fert. Soils,* 21, 84–88.

Haijun, L., Guanhua, H. (2009). Laboratory experiment on drip emitter clogging with fresh water and treated sewage effluent. *Agric. Water Manag.*, 96, 745-756.

Holt, J. G., Krieg, N. R., Sneath, P. H. A., Stanley, J. T. & Williams, S. T. (1994). In Williams & Wilkins (Eds.), Bergey's manual of determinative bacteriology (9th edition). Baltimore, Md, USA.

Hovanec, T. A., Taylor, L. T., Blakis, A. & Delong, E. F. (1997). Nitrospira-like bacteria associated with nitrite oxidation in freshwater aquaria. *Appl. Environ. Microbiol.*, 64, 258-264.

Jacobs, A. & Chenia, H. Y. (2011). Biofilm formation and adherence characteristics of an *Elizabethkingia meningoseptica* isolate from *Oreochromis mossambiscus.* Ann. Clin. Microbiol. Antimicrob., 10, 16.

Kreij, C., van der Burg A. M. M. & Runia, W. T. (2003). Drip irrigation emitter clogging in Dutch greenhouses as affected by methane and organic acids. *Agric. Water Manag.*, 60, 73-85.

Kuske, C. R., Barns, S. M. & Busch, J. D. (1997). Diverse uncultivated bacterial groups from soils of the arid southwestern United States that are present in many geographic regions. *Appl. Environ. Microbiol.*, 63, 3614–3621.

Laskin, A. I. & White, D. C. (1999). Preface to special issue on *Sphingomonas*. *J. Ind. Microbiol. Biotechnol.*, 23, 231.

Liu, Z., Huang, S., Sun, G., Xu, Z. & Xu, M. (2012). Phylogenetic diversity, composition and distribution of bacterioplankton community in the Dongjiang River, China. *FEMS Microbiol. Ecol.*, 80, 30-44.

Lymperopoulou, D. S., Kormas, K. A. & Karagouni, A. D. (2012). Variability of prokaryotic community structure in a drinking water reservoir (Marathonas, Greece). *Microbes Environ.*, 27, 1-8.

Lubello, C., Gori, R., Nicese, F. P. & Ferrini, F. (2004). Municipal-treated wastewater reuse for plant nurseries irrigation. *Water Res.,* 38, 2939-2947.

Magurran, A. E. (1988). Ecological diversity and its measurement. London: Princeton Univesity Press.

Muyzer, G., Brinkhoff, T., Nübel, U., Santegoeds, C., Schäfer, H. & Wawer, C. (1998). Denaturing gradient gel electrophoresis DGGE in microbial ecology. In A. D. L. Akkermans, J. D. van Elsas & F. J. Bruijn (Eds.), *Molecular microbial ecology manual* (pp. 1-27). Dordrecht: Kluwer.

Okami, Y. & Hotta, K. (1988). Actinomycetes in biotechnology: an overview. In M. Goodfellow, S. T. Williams & M. Mordarski. (Eds.), Actinomycetes in Biotechnology (pp. 37–67). London, UK: Academic Press.

Oron, G., Demalach, J., Hoffman, Z. & Manor, Y. (1996). Effect of effluent quality and application method on agricultural productivity and environmental control. *Water Sci. Technol.*, 7-8, 1593-1601.

Peczynska-Czoch, W. & Mordarski, M. (1988). Actinomycetes enzymes. In M. Goodfellow, S. T. Williams & M. Mordarski (Eds.), Actinomycetes in Biotechnology (pp. 219–284). London, UK: Academic Press.

Piret, J. M. &. Demain, A. L. (1988). Actinomycetes in biotechnology: an overview. In M. Goodfellow, S. T. Williams & M. Mordarski (Eds.), Actinomycetes in Biotechnology (pp. 461–482). London, UK: Academic Press.

Pitts, D.J. (1997). Evaluation of micro-irrigation systems. South West Florida Research and Education Center, University of Florida, USA, 46pp.

Puig-Bargués, J., Arbat, G., Elbana, M., Duran-Ros, M., Barragán, J., Ramírez de Cartagena, F. & Lamm, F. R. (2010). Effect of flushing irrrigation with effluents. *Agric. Water Manag.*, 97, 883-891.

Quaiser, A., Ochsenreiter, T., Lanz, C., Schuster, S. C., Treusch, A. H., Eck, J. & Schleper, C. (2003). Acidobacteria form a coherent but highly diverse group within the bacterial domain: evidence from environmental genomics. *Mol. Microbiol.* 50, 563–575.

Ravina, I., Paz, E., Sofer, Z., Marcu, A., Shisha, A., Sagi, C., Yechialy, Z., Lev, Y. (1997). Control of clogging in drip irrigation with stored treated municipal sewage effluent. *Agric. Water Manag.*, 33 (2-3), 127-137.

Sánchez, O., Gasol, J. M., Massana, R., Mas, J. & Pedrós-Alió, C. (2007). Comparison of different denaturing gradient gel electrophoresis primer sets for the study of marine bacterioplankton communities. *Appl. Environ. Microbiol.*, 73, 5962-5967.

Sánchez, O., Ferrera, I., Garrido, L., Gómez-Ramos, M. M., Fernández-Alba, A. R. & Mas, J. (2014). Prevalence of potentially thermophilic microorganisms in biofilms from greenhouse-enclosed drip irrigation systems. *Arch. Microbiol.*, 196, 219-226.

Schauer, M., Massana, R. & Pedrós-Alió, C. (2000). Spatial differences in bacterioplankton composition along the Catalan coast NW Mediterranean assessed by molecular fingerprinting. *FEMS Microbiol. Ecol.*, 33, 51-59.

Tarchitzky, J., Rimon, A., Kenig. E., Dosoretz, C. G. & Chen, Y. (2013). Biological and chemical fouling in drip irrigation systems utilizaing treated wastewater. *Irrig. Sci.*, 31, 1277-1288.

Taylor, H. D., Bastos, R. K., Pearson, H. W. & Mara, D. D. (1995). Drip irrigation with waste stabilization pond effluents, solving the problem of emitter fouling. *Water Sci. Technol.*, 31, 417-424.

Thierry, S., Macarie, H., Iizuka, T., Geissdorfer, W., Assih, E. A., Spanevello, M., Verhe, F., Thomas, P. Fudou, R., Monroy, O., Labat, M. & Ouattara, M. (2004). *Pseudoxanthomonas mexicana* sp. nov., and *Pseudoxant-homonas japonensis* sp. nov., isolated from diverse environments, and emended descriptions of the genus *Pseudoxanthomonas* Finkmann et al. 2000 and of its type species. *Int. J. Syst. Evol. Microbiol.,* 54, 2245-2255.

Vaz-Moreira, I., Egas, C., Nunes, O. C. & Manaia, C. M. (2011). Culture-dependent and culture-independent diversity surveys target different bacteria, a case study in a freshwater sample. *Antonie Van Leeuwenhoek*, 100, 245-257.

Yan, D., Bai, Z., Mike, R., Gu., L, Ren, S. & Yang, P. (2009). Biofilm structure and its influence on clogging in drip irrigation emitters distributing reclaimed wastewater. *J. Environ. Sci.*, 21, 834-841.

Zhang, J., Zhao, W., Wei, Z., Tang, Y. & Lu, B. (2007). Numerical and experimental study of hydraulic performance of emitters with arc labyrinths channels. *Comput. Electron. Agric.*, 56, 120-129.

Zucchi, T. D., Guidolin, A. S. & Cônsoli, F. L. (2011). Isolation and characterization of actinobacteria ectosymbionts from *Acromyrmex subterraneus brunneus* (Hymenoptera, Formicidae). *Microbiological Res.,* 166, 68–76.

In: Drip Irrigation
Editor: Alfred H. Steele
ISBN: 978-1-63463-737-4

Chapter 3

ABSORBENT POLYMERS AND DRIP IRRIGATION: A WAY OF CONTROLLING IRRIGATION WATER

Mohamed R. Berber[1,2,*] ***and Inas H. Hafez***[1,3]
[1]Department of Applied Chemistry, Graduate School of Engineering, Kyushu University, Fukuoka, Japan
[2]Department of Chemistry, Faculty of Science, Tanta University, Tanta, Egypt
[3]Department of Natural resources and Agricultural Engineering, Faculty of Agriculture, Damanhour University, Damanhour, Egypt

ABSTRACT

With no doubt, polymers applications have a great influence on the way of our life because of their ease of processing that meet the specific needs. In agriculture, polymers are widely used to protect the environment using new techniques of applications. In particular, the polymeric materials are used into agricultural soils to improve aeration, conserve moisture, provide mulch, and to promote the fertility and the health of the soil. Consequently, polymers are receiving a great deal of attention as soil conditioners. Here, we focus on soil-polymer applications under surface- and subsurface-drip irrigation systems to manage the irrigation water and to improve the soil properties such as stability and water holding capacity. In this chapter, we highlight the main advantages and the drawbacks of surface- and subsurface-drip irrigation systems that use polymer conditioners in view of system construction, quality of irrigation water, evaporation and percolation, crop production, environmental regulation and safety. We also discuss in details the effects of polymer type (synthetic and natural), polymer concentration and polymer formulation process on soil properties, in view of water-holding capacity, water use efficiency, soil permeability and infiltration rates, and plant performance especially in structureless soils. In addition, we illustrate the effects of soils and water properties on polymer absorbency of water. Moreover, we show the synthesis and the characterization of the different absorbent polymeric materials and how parameters such as initiator type, initiator concentration, and polymerization type affect the polymer capacity of water

* E-mail: mrberber@mail.cstm.kyushu-u.ac.jp; mrberber@science.tanta.edu.eg. Tel/Fax: +81928022845.

absorptivity. Subsequently, we throw some light on the future impacts of the polymeric materials in the agricultuel area.

1. INTRODUCTION

Although over two thirds of Earth's surface is covered by water, more than 97% is ocean water which is too salty for human use or even for irrigation, consequently, the freshwater is only 3%. Almost, 1% of the available fresh water is liquid surface water that can be used directly by human and the rest is ground water and iced water. [1] Accordingly, fresh water is considered as one of the most abundant resources on earth.

In the agriculture sector only, more than two-thirds of the available freshwater is used for irrigation. [2] In this context, water used for irrigation is receiving a great deal of concern. Figure 1 illustrates the water content on earth and its distribution.

2. IRRIGATION SYSTEMS

In order to manage the irrigation water, different irrigation systems, e.g. surface, sprinkler, trickle and sub-surface drip irrigations were established depending on the soil type, climate, water quality and availability. [3-8] For example, in clay soils which have low infiltration rates, surface irrigation is commonly applied. While, in sandy soils which have low water holding capacity and high infiltration rates, drip irrigation are more preferred than surface irrigation. The climate also play an important role is the selection of the irrigation system. For example, in windy climate, surface or subsurface-drip irrigations are preferred than sprinkler irrigation where the wind can disturb the spraying of water from sprinklers.

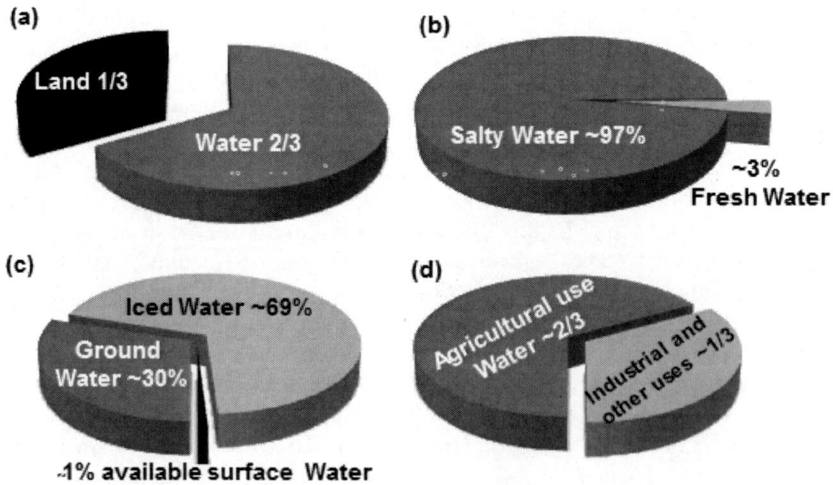

Figure 1. (a) Earth total water, (b) Fresh Water content, (c) Fresh Water distribution on earth. (d) Fresh water uses on earth.

Also, water quality has an essential role in the selection process of the irrigation system. Surface irrigation is preferred if the irrigation water contains much sediment, where the sediments can block the drippers or the sprinklers of the drip and the sprinkler irrigation systems, respectively. Furthermore, the application of soil conditioners determines the choice of the irrigation system. If the irrigation water contains dissolved salts, drip irrigation is particularly suitable, as less water is applied to the soil. The subsurface drip irrigation also is more efficient than surface irrigation in the case of using leached salts into the irrigation water. Accordingly, the choice of the irrigation system depends on many factors.

In this chapter, we mainly focus on the management of irrigation water in sandy soils by using polymers as soil conditioners and by applying the drip irrigation systems.

3. Sandy Soils and Drip Irrigation Systems

3.1. Sandy Soils

Sandy soil is mainly composed of sand and smaller amounts of silt and clay compared to the clay soil which mostly contains silt and clay. [9] The water holding capacity of the sandy-soil is very small because the soil-particles size is too large. Hence, the irrigation water and the nutrients are quickly drained away from the plant root zone. Figure 2 shows the water holding characteristics of sandy soil in comparison with clay soil.

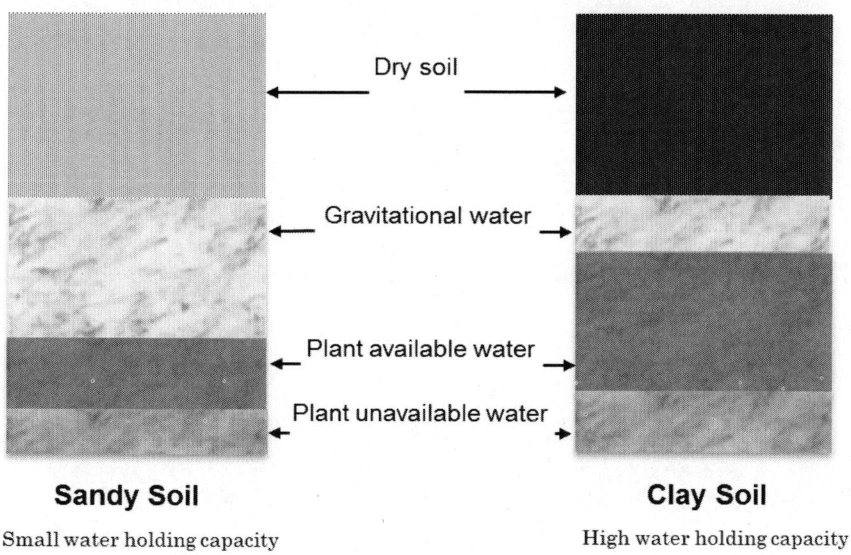

Figure 2. Water holding characteristics of sandy soil in comparison with clay soil.

3.2. Drip Irrigation

Drip irrigation is an irrigation system that applied water to soils through mechanical devices called emitters located at points along the water delivery lines. The emitters dissipate the pressure from the distribution system by means of orifice vortexes and tortuous, thus allowing a limited volume of water to discharge. The emitters are commonly placed on the ground (surface drip irrigation) or under the ground (subsurface drip irrigation) in the soil at shallow depth. [5] Drip irrigation has a high potential in reducing energy use and water and soluble nutrient losses. The uniformity of the drip irrigation systems strongly depend on emitter manufacturing variation, land slope-induced hydraulic variability of the irrigation unit and head losses in pipes, emitter sensitivity to pressure and temperature variations, and emitter clogging.

3.3. Subsurface Drip Irrigation in Comparison with Surface Drip Irrigation

Subsurface drip irrigation system (SDI) is a low-pressure and a high efficiency irrigation system that uses buried plastic tubes containing embedded emitters located at a regular spacing, in order to improve the irrigation-water use efficiency through the reduction in water losses from the plant rooting zone, where adequate storage of moisture and nutrients are required for optimizing crop production (Figure 3).

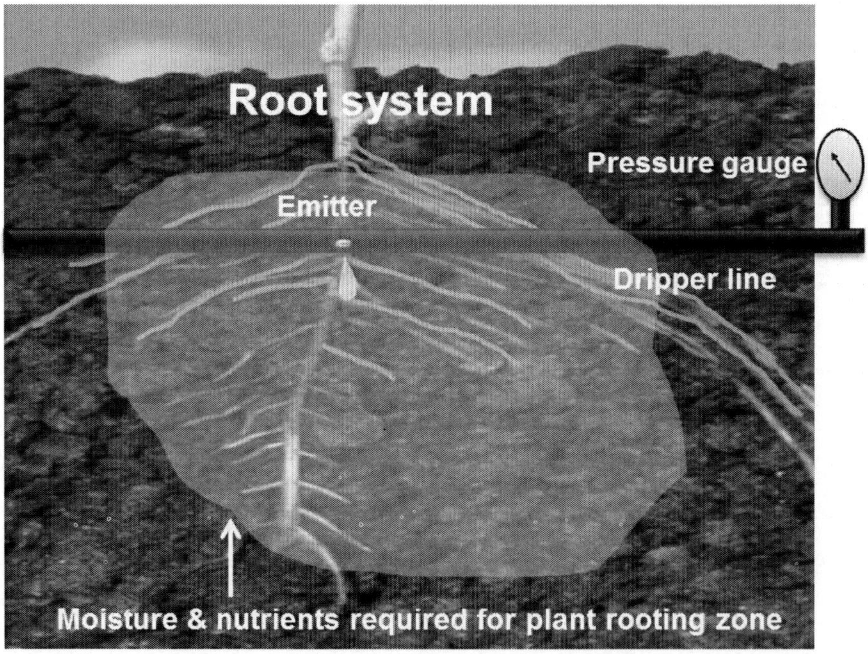

Figure 3. Illustration of water distribution at plant zone using sub-surface drip irrigation system.

SDI provides the ultimate in water use efficiency for open-field agriculture, often resulting in water savings of 25-50% compared to surface irrigation. The design of the SDI system is the same as for surface drip irrigation system except the tubing is buried. Burying the tubing adds additional initial cost to the system but eliminates the need to install and remove tubing at the beginning and end of each growing season. Because the SDI system is no longer in view, root intrusion, distribution uniformity, tubing damage from equipment and burrowing animals are all advantages over the surface drip irrigation system. [10] The use of

SDI also offers many other advantages for crop production, including less nutrients leaching compared to surface irrigation, higher yields, a dry soil surface for improved weed control and crop health, the ability to apply water and nutrients to the most active part of the root zone, and the ability to safely irrigate with wastewater while preventing human contact. [11, 12] One of the key differences between SDI and surface drip irrigation is the water movement at the emitter point. In surface irrigation, a semicircle region is formed under the emitter source, while a spherical-shaped saturated region is formed around the emitterof the SDI. [13].

4. Polymers As Soil Conditioners

A soil conditioner is a material that can be added into soil to improve its physical properties, in particular its water retention and penetration, its drainage and its ability to retain nutrients. A wide variety of materials have been described as soil conditioners due to their ability to improve soil properties, e.g. polymers.

With no doubt the use of polymer-based products for water clarification and control is receiving a great deal of attention. [14-16] Polymers are natural or synthetic organic materials composed of many small molecules (monomers) that are arranged in a simple repeating structural design to form a large molecule. There are several categories of polymers depending on the polymer architecture and shape (microscale ordering of the monomers within the polymer chain), and the definition of the polymer physical and chemical properties, as well as the polymer application target. Polymers are characterized by different physical and chemical properties, e.g. weight average molecular weight, the number average molecular weight, polydispersity (the broadness of molecular weight distribution), polymer conformation (the space occupied by a polymer molecule), polymer crystallinity (intramolecular folding and stacking of the adjacent chains in the three-dimension) and Polymer charge and its density. These structural characteristics play an essential role in determining how the polymer behaves as a continuous macroscopic material, and how the polymer chains interact with the other organic compounds through the various forces (e.g. ionic, covalent and non-covalent forces). [17]

In agriculture, the early use of polymers was as mulches and shelters depending on polymers transparency, stability and permeability. [18] These applications have considered polymers as inert materials rather than as active compounds. However, with the innovations of the functionalized polymers and the revolution of polymer industry, polymers have found different and versatile agricultural applications. Polymers are considered as a type of soil conditioners that can be used to improve plant performance in arid soils, water use efficiency, soils water holding capacity, soil permeability and infiltration rates, in addition to, the reduction of erosion and water run-off, irrigation frequency, and compaction tendency. They are also used to adjust the soil pH to meet the needs of specific plants or to make highly acidic or alkaline soils more applicable. [19]

4.1. Synthetic Organic Polymers and Irrigation

Synthetic polymers are very popular in soil applications of water retention due to their ability to absorb water several hundred times of their own weight and their stability towards environmental breakdown. [20] Polyacrylonitrile (Figure 4a) polymer was introduced as the first synthetic soil conditioner that can improve the soil water holding capacity. [18] Since that time, many different polymers were introduced to the soil-water applications as superabsorbent polymers, e.g. polyacrylamide (PAAm; Figure 4b) and polyacrylate (PAA; Figure 4c).

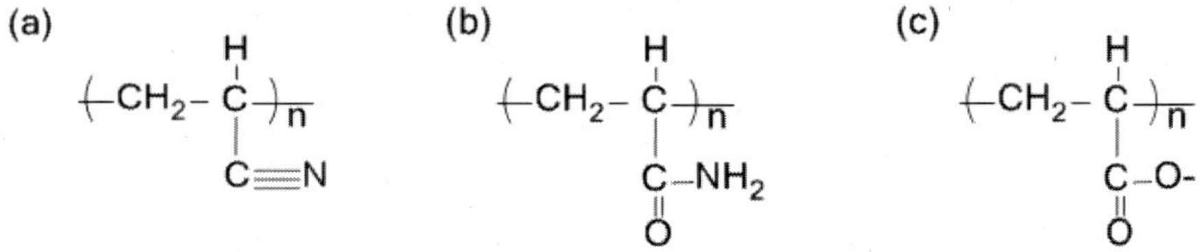

Figure 4. The chemical structure of: Polyacrylonitrile (a), PAAm (b), and PAA (c).

4.1.1. PAAm and Irrigation

PAAm is a polymer prepared from acrylamide monomers. It can be synthesized as a simple linear-chain or cross-linked polymer. PAAm is a not toxic material. In the crosslinked form, the possibility of the monomer being presentwithout polymerization is reduced even further. PAAm is a highly water-absorbent polymer, forming a soft gel when hydrated. It was tested as a soil conditioner for improving soil properties and water availability, especially in sandy soils. PAAm is attracted to soil particles via columbic and Van der Waals forces. [21] These surface attractions stabilize soil structure by enhancing particle cohesion, thus increasing resistance to shear-induced detachment and preventing transport in runoff.

Hafez et al. [22] showed the effect of different application parameters of PAAm on water absorbency using subsurface drip irrigation system, e,g, dripper line depth, operating water discharge rate and polymer concentration (0, 0.01, 0.05 and 0. 1 wt%) at 60 cm soil depth. The irrigation experiments were conducted in a wooden soil box with a transparent Plexiglass side by using a subsurface drip irrigation system. The soil box was designed to simulate a soil section of the field. The soil-moisture content (MC) was determined inside the formed water bulb under and above the water emission point in three dimensions. The MC ratio was measured as follows. Wetted samples of known weight (W_w) were placed into a porcelain crucible, and then were dried in an oven at 105 °C for about 24 h. After removing the samples from the oven, they were slowly cooled in a dissector at room temperature and reweighed again (W_d). The MC% was determined by using the following equation. $MC\% = [(W_w-W_d)/W_d]*100$. The results indicated that the shallower the dripper line depth the longer the time taken to reach the maximum soil depth. Figures 5 and 6 show the effect of different operating pressures (1.0, 1.5 and 2.0 bar) on the waterfront advance distributions at dripper line depth 20 cm and at different polymer concentrations. The results also showed indicated that the lower operating pressure, the better the water distribution, especially above the dripper line. Furthermore, the contour lines were symmetrical around the dripper line under the operating pressure (1.0 bar). The time of the waterfront progress to reach the soil box depth was longer with the lower operating pressure (1.0 bar). Figure 7 shows the effect of the polymer concnetration on the average moisture distribution under the emitter at dripper line depth 20 cm and operating pressure 1.0 bar. As observed from the figure, the increase of the polymer concentration has led to a longer time of the waterfront to reach the soil box depth. In other words, an increase in the water holding capacity was achieved with the increase of the polymer concentration.

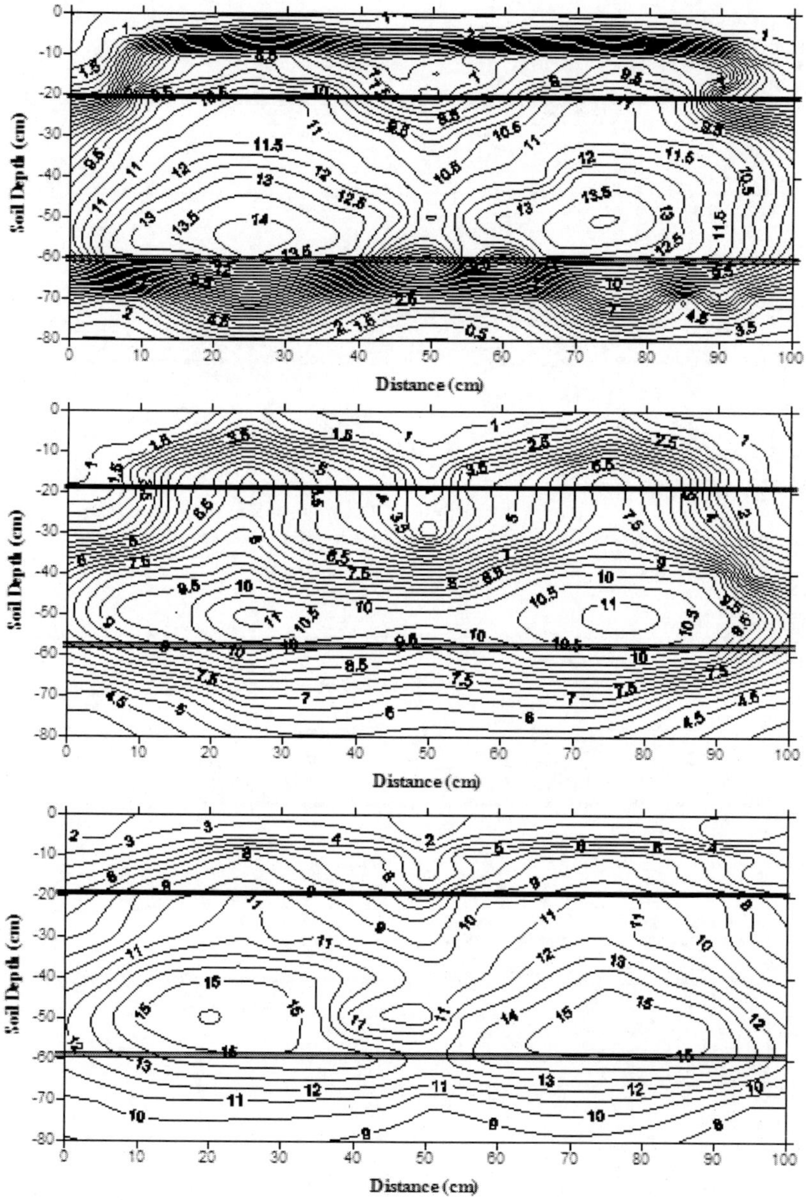

Figure 5. The moisture distribution through two emitters into the soil box under different operating pressure for dripper line depth 20 cm and polymer concentration 0.01 wt%. Operating pressure 1.0 bar (upper panel), operating pressure 1.5 bar (middle panel), and operating pressure 2 bar (lower panel).

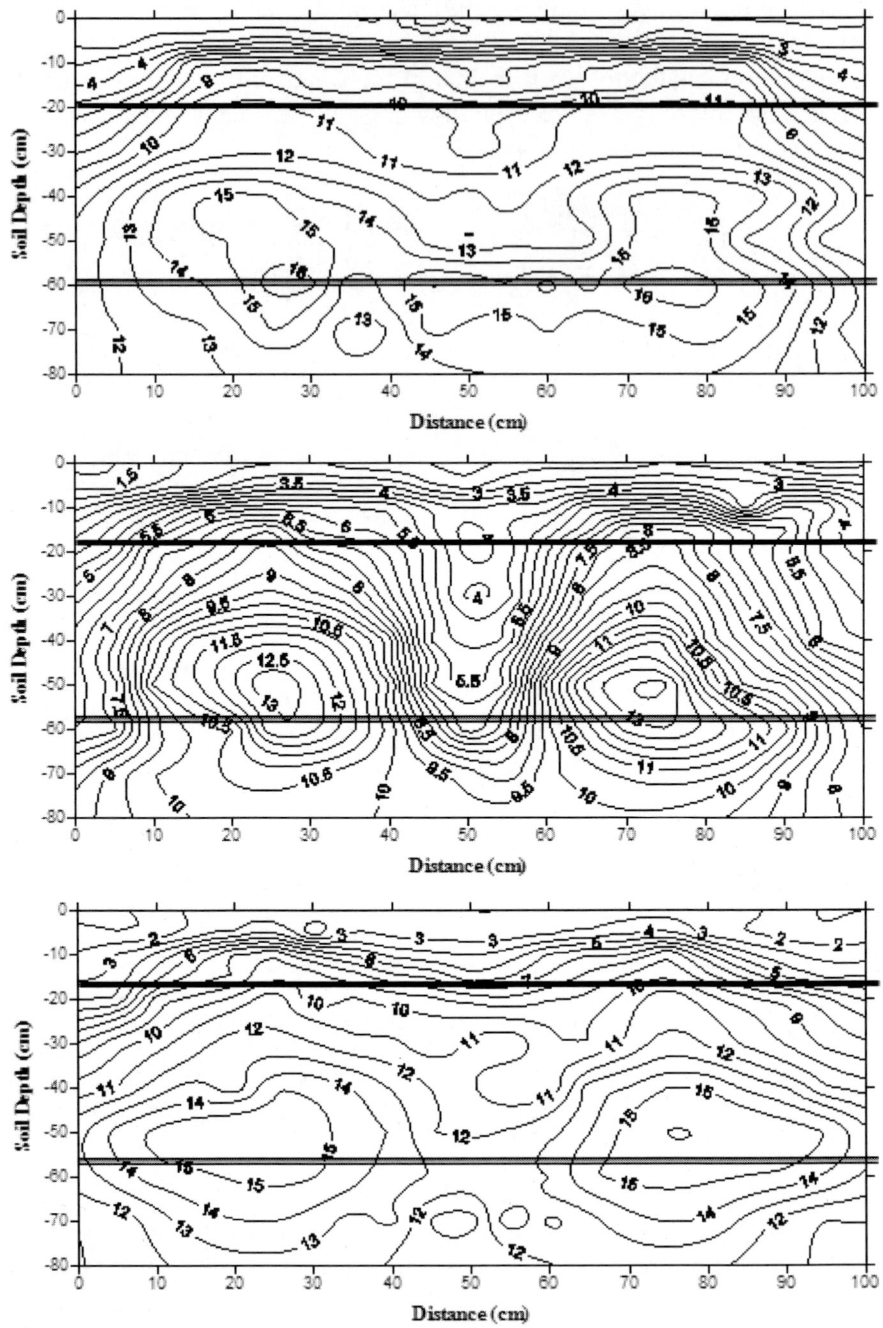

Figure 6. The moisture distribution through two emitters into the soil box under different operating pressure for dripper line depth 20 cm and polymer conc. 0.1wt %. Operating pressure 1.0 bar (upper panel), operating pressure 1.5 bar (middle panel), and operating pressure 2 bar (lower panel).

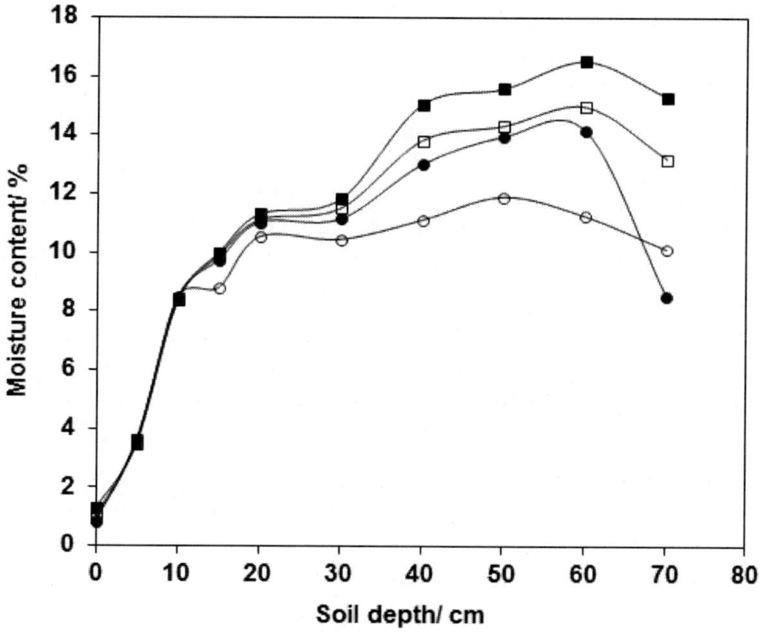

Figure 7. The effect of polymer layer concentrations on the average moisture distribution under the emitter at dripper line depth 20 cm and operating pressure1.0 bar. Control without polymer (○), polymer concentration 0.01wt% (●), polymer concentration 0.05wt% (□), and polymer concentration 0.1wt% (■).

The water-saving and yield-increasing effect of PAAm was also studied through a plot experiment. The results of this study showed that soil water content increased with the increase of the application concentration of PAAm. During the whole growth period of maize, soil water content increased by 5.62%~10.96% in comparison with control. After the application of PAAm, as for all the treatments the average plant heights of maize were higher than control; ear height, ear length and ear diameter had little change; a great effect on the bald length of maize. Meanwhile, PAAm did not have an adverse effect on photosynthesis of maize leaves, transpiration, leaf temperature and crop water use efficiency. This study also showed that within the application concentration of 0.5~2.0 g/m^2 of PAAm, the maize yield increased by the percentages of 5.54%~14.13%. [23] PAAm incorporation into soil has also led to a better aeration for the root system, a reduction of soil compaction and an increase in both the nutrient and water supply to the roots. [24]

PAAm copolymers-based hydrogels were also used to examine the physico-bio-chemical properties of a sandy soil under drip irrigation. In one study, the examined application rates of the PAAm-based hydrogel were 2, 3 and 4 g/plant pit. The obtained results were summarized as follows: 1-the applied PAAm hydrogel positively affects the hydrophysical properties of the soil. These include; a- improving soil structure expressed by water stable structural units > 0.25 mm in diameter and structure coefficient, dry stable structural units > 0.84 mm in diameter and wind erosion parameter indicating high resistance of the soil against both wind and water erosion and the destruction of the soil by tillage operations. b- decreasing soil bulk density as well as macro porosity (drainable pores) on the expense of micro ones. Therefore, water holding pores were increased. c- decreasing mean diameter of soil pores and in turn its water transmitting properties namely: infiltration rate, and hydraulic conductivity for vertical flow of water through soil profile. Evaporation was also decreased. 2-soil conditioning with PAAm hydrogel positively affected the chemical and the biological properties of the soil. These effects are assembled in the following; a- the pH of the soil was slightly decreased. b- increasing both conductivity of the soil and its specific surface area, indicating an improvement in activating chemical reactions in the soil. c- increasing of organic carbon and total nitrogen in the soil. d- improving the biological activity expressed as total count of bacteria, fungi and actinomycetes in the soil. [25]

In the same regard, the effect of a crosslinked-type PAAm, on water-holding capacity of a sandy soil was studied under the laboratory and glasshouse conditions. Water-holding capacity of the soil exposed to 0.01 MPa increased by 23% and 95% by adding 0.03% and 0.07% of PAAm to the soil, respectively. This indicated that the soil treated with PAAm was able to store more water than untreated soil, thereby reducing the potential losses due to deep percolation in sandy soils. However, the PAAm in the treated soil did not significantly increase the quantity of water released from the soil by increasing the pressure from 0.01 to 1.5 MPa.

The results from the first glasshouse experiment demonstrated that the excess amount of water stored in the soil by PAAm was available to plants and resulted in higher water use and grain production.

Consequently, there was a 12 and 18 fold increase in water use efficiency of soybean plants grown in soils treated with 0.03% and 0.07% PAAm, respectively. The results from the second glasshouse experiment demonstrated that increasing amounts of PAAm in a sandy soil can extend the irrigation interval without any adverse effect on the grain yield of soybeans. [26].

4.1.2. Paam Water Absorbency

PAAm has a water absorptivity of around 400 times its dry weight. Several trials were done to highly improve the water absorbency of PAAm. For example, high water-absorbent copolymers comprising acrylamide and acrylic acid were prepared in the presence of a crosslinking agent by a solution polymerization technique using a redox initiation system. Such copolymers showed very high water absorbency and absorbing kinetics to the distilled water. The maximum water absorbency recorded from this hydrogel type was about 900 g-water/g-dry copolymer. [27]

In another trial, PAAm was crosslinked with 2,3-dihydroxybutanedioic acid (DBA) in aqueous medium via gamma-radiation. The influence of gamma absorbed dose and DBA content of the hydrogels on the swelling properties were examined. The obtained PAAm hydrogels showed enormous swelling in an aqueous medium and displayed swelling characteristics which were highly dependent on the chemical composition of the hydrogels and irradiation dose. The increase of the gamma-radiation dose decreased the gel hydrophilicity, leading to a lower swelling ratio. On the other side, the increase of the DBA content in the copolymers has led to an increase in the water absorption. The values of the weight swelling ratio of obtained hydrogels were between 8.34% and 15.16%, while the values of the weight swelling ratio of pure PAAm hydrogels were between 7.58% and 8.28%. Water diffusion into hydrogels was found to be non-Fickian in character. [28]

PAAm was also crosslinked with Poly(vinyl alcohol) (PVA) which is a highly hydrophilic polymer in order to improve the PAAm water absorbency. [29] Different synthetic parameters such as PVA/acrylamide ratio, initiator and solvent types were applied to determine the effects of the synthetic conditions on the gel properties. The gel prepared with PVPA/PAAm ratio of 2,33/1 using potassium persulfate/ferrous sulfate as an initiator in H_2O solvent showed an equilibrium water content of about 70-71% after 2-3 h immersion in distilled water at 25°C. [30] Figure 8 summarizes the water absorbency of PAAm-based hydrogels.

4.1.3. Effects of Soils and Water Properties on Polymer Absorbency

Soil and water properties such as texture, clay mineralogy, organic matter and concentration of dissolved salts were also reported to affect the absorptivity of the polymer. The sorption isotherms of several types of poly anion polymer were determined in sandy loam soils. [31, 32] The results suggested that polymer sorption by soil was mostly limited to the external surface area and was considerably influenced by water quality. [33] In addition, there was a little desorption after the polymers were adsorbed onto soil. [34]

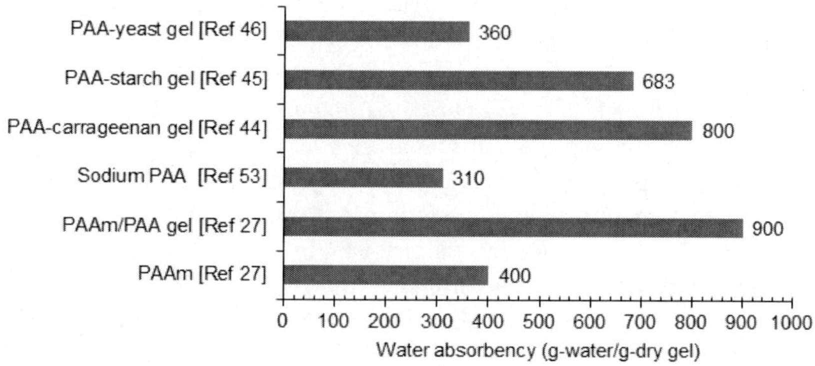

Figure 8. Water absorbency of PAAm-based and PAA-based hydrogels.

For example, soils with high clay or silt content and low organic matter content were reported to have a high sorption affinity for anionic PAAm. The amount of PAAm sorption was found to significantly increase as the total dissolved salts increased. The divalent cations Ca^{2+} and Mg^{2+} were about 28 times more effective in enhancing PAAm sorption into soils than the monovalent cations Na^+ and K^+, mainly because of their stronger charge screening ability. The effectiveness of cation enhancement on PAAm sorption varied with soil texture and was greater in fine soils than in sandy soils. The organic matter content was found to have a negative effect on PAAm sorption. Soil samples after the removal of partial organic matter adsorbed more PAAm than natural soils. The negative effect of organic matter on PAAm sorption was attributed to the reduction of accessible sorption sites by cementing inorganic soil components to form aggregates and to the enhancement of electrostatic repulsion between PAAm and soil surface by its negatively charged functional groups. [35]

4.1.4. PAA and Soil Applications

PAA are a family of anionic polyelectrolyte vinyl polymers with the ability to absorb and retain water and also swell many times their original volume. There are different types of

PAA e.g. sodium PPA and potassium PAA which are the most widely used polymers today. PAA has different agricultral applications including soil remediation, plant seedling performance and irrigation water management. [36]

About soil remediation, PAA were used to improve the soil quality and to reduce the contamination of the soil. PAA were able to reduce heavy metal and salt stress to crops by accumulating the soil inorganic compounds. [37, 38] In many studies, the effect of sodium PAA on the physical properties of different soils e.g. sand, peat and chernozem was determined. For example, the application of sodium PAA in a concentration of 0.5wt% in a sandy soil has led to a remarkable change in the soil water absorbency and the soil physical properties. Also, the amount of plant-available water and water retention at different pressures (0-1.5 MPa) was increased, whereas the bulk density was decreased. Furthermore, significant water saving was achieved when the PAA hydrogels are applied to sandy soils. The PAA hydrogel application showed no change in the irrigation water needs of organic soils with higher content of clays, and also the aboveground biomass production in these soils remained unchanged. [39]

Regarding the plant seedling, potassium PAA was used to enhance the plant-available water in poor arable soils. The effect of polyacrylate hydrogel on plant growth during non-water stress conditions is examined through the root and shoot biomass of seedlings of nine tree species; *Eucalyptus grandis*, *Eucalyptus citriodora*, *Pinus caribaea*, *Araucaria cunninghamii*, *Melia volkensii*, *Grevillea robusta*, *Azadirachta indica*, *Maesopsis eminii* and *Terminalia superba*. The seedlings were potted in five soil types; sandy, sandy loam, loam, silt loam and clay. These pots were amended at two hydrogel levels: 0.2% and 0.4% w/w and grown under controlled conditions in a green house. Root and shoot growth responses of the seedlings were determined by measuring the dry weight of the roots, stems, leaves and twigs. The addition of either 0.2% or 0.4% hydrogel to the five soil types resulted in a significant increase of the root dry weight in eight tree species compared to the controls after 8 weeks of routine watering. Also, the dry weight of stems, leaves and twigs were significantly higher in the nine tree species potted in hydrogel amended soil types than in the hydrogel free controls. These results suggested that PAA hydrogel amendment enhances the efficiency of water uptake and utilization of photosynthates of plants grown in soils which have water contents close to field capacity. [40] The effect of cross-linked PAA hydrogel amendment on the performance of tree seedlings of Picea abies, Pinus sylivestris and Fagus sylvatica grown in temperate soils under water stress and non-water stress periods was also investigated in a green house. The objective was to compare the root and shoot biomass of seedlings of the three species grown in sand, loam and clay soils amended with 0.4% w/w hydrogel in non water stress conditions as well as survival, root and shoot biomass after subjection to water stress. The seedlings were grown for 16 weeks, harvested and shoot as well as root biomass determined before water stress. The seedlings were also subjected to water stress and their biomass assessed at death following the water stress. The results showed that root and shoot biomass were generally higher in hydrogel amended soils compared to the controls. Root and shoot biomass of Fagus sylvatica was lower compared to Picea abies and Pinus sylivestris before water stress. The 0.4% hydrogel amendment significantly increased species' survival in the different studied soils. Although root biomass was higher in hydrogel amended sandy soil compared to other soils, P. sylivestris and F. sylvatica shoot biomass were higher in hydrogel amended clay and loam soils compared to the sandy soil after water stress. Biomass was higher in sandy soil compared to loam and clay soils under non-water and water stressed conditions. Based on these studies, the use of PAA hydrogels could be promoted to enhance seedling production in water stress and non-water stress environments. [41]

In the case of irrigation water management, the effect of sodium PAA on water retention, saturated hydraulic conductivity (Ks), infiltration characteristic and water distribution profiles of a sandy soil was investigated at different application rates. The results showed that water retention and available water capacity effectively increased with increasing PAA concentration. The Ks and the rate of wetting front advance and infiltration under certain pond infiltration was significantly reduced by increasing PAA concentration, which effectively reduced water in a sandy soil leaking to a deeper layer under the plough layer. The effect of PAA on water distribution was obvious to the upper soil layers and very little to the following deeper layers. [42]

4.1.5. Water Absorption Efficiency of PAA

Different synthetic parameters were proved to significantly affect the PAA water absorption capacity e.g. polymerization technique, polymerization conditions such as monomer concentration, pH of the medium and crosslinker type and its amount. [43] A series of superabsorbent PAAm hydrogels were prepared from carrageenan and partially neutralized acrylic acid by gamma irradiation at room temperature. The gel fraction, swelling kinetics and the equilibrium degree of swelling (EDS) of the hydrogels were studied in details. Under optimum conditions, a PAAm-carrageenan hydrogel was prepared with a very high swelling ratio (~800 g-water/g-dry gel). This hydrogel was also found to be sensitive to the pH and the ionic strength of the medium. [44] Also, PAA-based starch superabsorbent hydrogel was simply prepared by free-radical graft polymerization of acrylic acid and potato starch in the aqueous solution by using Triton X-100 as the pore-forming agent. The hydrogel prepared with an acrylic acid/starch ratio of 6:1 and a Triton X-100 content of 0.4 wt% showed a water absorbency of 683 g-water/g-dry gel in distilled water. [45] Recently, acrylic acid-based yeast copolymers were successfully synthesized by graft copolymerization of acrylic acid onto the surface of yeasts by using ammonium persulfate as a free-radical initiator and N,N'-methylene-bisacrylamide as a crosslinker in aqueous solution. The weight ratio of acrylic acid to yeast was investigated. The results showed that the water absorption capacity of the yeast in distilled water was enhanced with the increase in the mass ratio of acrylic acid up to 8% (~360 g-water/g-dry gel). Furthermore, six-time consecutive adsorption−dsorptiona cycles, which are promising for the potential applications in agricultural areas were recorded. [46] Figure 8 displays the water absorbency of the discussed PAA-based hydrogels.

4.2. Natural Polymers and Irrigation

Natural polymers occur in nature and can be extracted from natural materials. They are often water-based and usually formed through a condensation polymerization process. They can be used to retain moisture in the surrounding soil during plant growth and transportation, they can also act as water reservoirs in soil, releasing water into the soil and maintaining the soil moisture balance. [47]

The great advantages of natural polymers over synthetic polymers in irrigation are their biodegradability, non-toxicity and shear resistance characteristics.

4.2.1. Cellulose and Its Derivatives As Absorbent Polymers

Natural polymers such as cellulose (Figure 9) and its derivatives are receiving considerable attentions as water absorbent polymers because of their biodegradable characteristics, their natural abundance and their potential water absorptivity. The most common form of cellulose derivatives as an absorbent polymer is carboxymethylcellulose (CMC). CMC absorbs the fluids forming a gel material that completely has different physical properties compared to the dry state. The pH of the medium and the solution salinity strongly affect the absorptivity of CMC polymer. In this regard, the acidic mediums and the concentrated salt solutions decrease the absorption capacity of CMC, where the salt ions block the acidic group of CMC, leading to lower water absorbency.

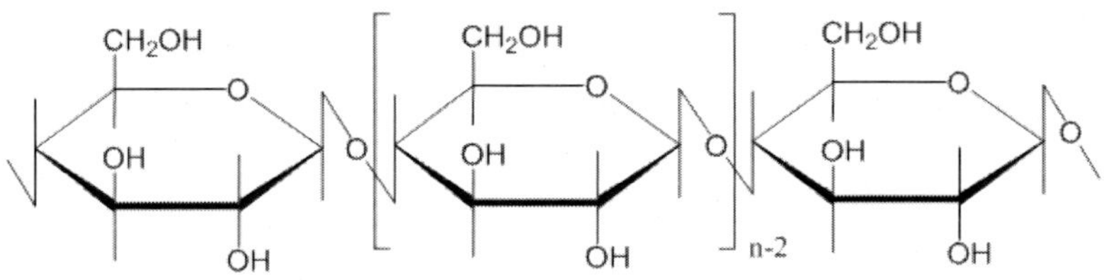

Figure 9. The chemical structure of cellulose.

The natural form of cellulose has a low water absorption capacity, however after chemical modification it shows a good absorptivity. The sulfonation of two or more hydroxyl groups of cellulose backbone is considered as one of the solutions to increase cellulose water absorbency. Since this process requires the substitution of two different hydroxyl groups. Thus, two different processing steps are required, leading to a higher production cost. In addition, the sulfonation process strongly affects the biodegradability. Hence, searching for new techniques of water absorptivity enhancement of cellulose was a market demand.

Grafting of cellulose and its derivatives with other polymers is a way of improving the water absorption capacity. It is well known that cellulose-based hydrogels require a chemically stable and a biologically safe crosslinked network that can provide effective swelling and absorption properties. Divinylsulphone (DVS) was used as a crosslinking agent for a mixture of CMC sodium salt (CMCNa) with hydroxyethylcellulose (HEC). Remarkable differences in water sorption capabilities were detected by changing polymers composition and the content of the crosslinking agent. The highest water absorbency (450 g-water/g-dry gel) was recorded when the hydrogel is prepared by using CMCNa/HEC with a weight ratio of 3:1 (total polymer concentration 2 wt%) and DVS concentration of 0.04 mol/L. [48] In a recent work, cellulose fiber was crosslinked by esterification with poly(vinyl methyl ether-co-maleic acid) (PVMEMA) and polyethylene glycol (PEG) to improve its water absorbency. The effects of the cellulose fiber length, crosslinking reaction time, and dosage of PVMEMA on water absorption and retention values were studied in details.

The results showed that as the fiber length is mechanically decreased from 2.41 to 0.50 mm and employing a weight ratio of fiber to polymers equivalent to 1.00:1.28, the water absorption increased from 86.50 to 189.20 g-water/g-dry gel, which is 119% higher than does the original fiber-based hydrogel. After curing, the water absorption and retention value of the hydrogels increased dramatically by 1640, 2822, and 4853% compared to uncrosslinked fibers with fiber lengths of 2.41, 0.97 and 0.50 mm, respectively. [49] However, the compounds used as crosslinking agents are often highly toxic. [50] Thus, they are not recommended as starting materials in view of an environmentally safe production process. Hence, searching for a non-toxic water-soluble crosslinking agent was an environmental requirement.

Carbodiimide which is a non-toxic water-soluble compound [51] was used as a crosslinking agent to fabricate an environmentally superabsorbent cellulose-based hydrogel, using HEC and CMCNa as starting polymeric materials. [52] This cellulose-based hydrogel showed a remarkable increase in the equilibrium water uptake (425±27 g-water/g-dry gel) when the starting materials weight ratio was 1:1 of CMCNa/HEC (total polymer concentration 5 wt%) and carbodiimide concentration was 3 wt% of the polymer concentration. [52] Also, to overcome toxicity and cost production associated with other crosslinking reagents, citric acid (CA) which is a nontoxic, commercial and water soluble compound was used as a crosslinking agent to prepare a new environmentally and friendly superabsorbent hydrogels derived from cellulose materials. [53] The crosslinking of CMCNa and HEC (weight ratio 3:1 mixture) with a low-CA concentration (1.75% w/w) has provided a hydrogel with a a water absorbency of 1100 g-water/g-dry gel. [53]

Cellulose-based hydrogel prepared without a crosslinking agent was another way of improving the water absorbency of cellulose. A novel superabsorbent hydrogel was prepared by the reaction of cellulose with succinic anhydride in the presence of 4-dimethylaminopyridine as an esterification catalyst and in a solvent mixture of lithium chloride and N-methyl-2-pyrrolidinone, followed by sodium hydroxide neutralization. The obtained hydrogel absorbed an amount of water equal 400 times its dry weight (400 g-water/g-dry gel), and this was comparable to a conventional sodium polyacrylate superabsorbent hydrogel. This hydrogel biologically degraded almost after 25 days, showing an excellent biodegradability. [54]

Cellulose-based hydrogels is also prepared by ionizing irradiation technique. Biodegradable hydrogel of CMCNa was synthesized by the irradiation of the sample with gamma rays generated from a cobalt source at a dose rate of 10 kGy/h. The swelling ratio of the formed hydrogel was found to decrease with the increase of the irradiation dose. This is because of the increase in crosslinking density of the hydrogel. The CMCNa hydrogel showed a swelling of 165 g- water/g-dry gel after an irradiation dose of 5 kGy. [55] By the optimization of the irradiation dose and the polymer concentration, up to 95% of the polymer was transformed into a hydrogel with a water absorbency of 800 g of water per gram of dry gel. [56] Figure 10 summarizes the water absorbency of cellulose-based hydrogels.

The effects of CMC on the water holding capacity and the saturated hydraulic conductivity of a sandy soil at various soil temperatures and water quality were studied. The soil-carboxymethylcellulose mixtures exhibited different water retention characteristics under different soil temperature conditions. Also, it was found that available water content in the soil increased up to four times with CMC as compared to control soil. [57]

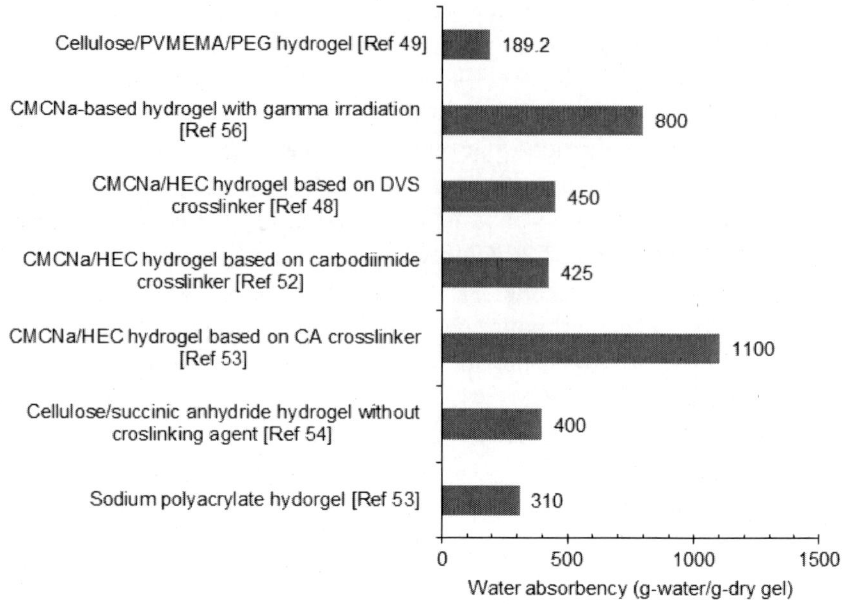

Figure 10. Water absorbency of cellulose-based polymers.

4.2.2. Starch As an Absorbent Polymer

Starch is a biodegradable and renewable natural polymer, containing a large number of hydrophilic hydroxyl groups in their backbone structure (Figure 11). The hydrophilicity of these hydroxyl groups makes starch a good material for absorbency. The molecular structure of starch is similar to that of cellulose. They are both made from glucose monomer and have the same glucose-based repeat units. The only difference in their structure is, in starch, all the glucose repeat units are oriented in the same direction, however in cellulose, and each glucose unit is rotated 180 degrees around the axis of the polymer backbone, relative to the last repeat unit.

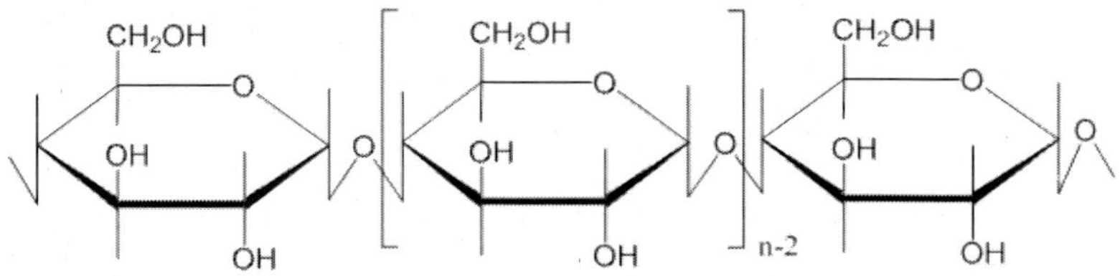

Figure 11. The chemical structure of starch.

Starch also has attracted a huge deal of attention similar to cellulose. Starch copolymers and hydrogels have offered remarkable improvements in the water absorption capacity and the soil physical, chemical, and biological properties, in addition to the plant growth parameters. [58] For example, a starch hydrogel was obtained by the reaction of starch with succinic anhydride (SA) in the presence 4-dimethylaminopyridine as an esterification catalyst in dimethylsulfoxide or water as a solvent, followed by NaOH neutralization. The water absorbency of the starch hydrogel obtained from dimethylsulfoxide solvent was about 120 g-water/g-dry gel, whereas the water absorbency of the starch hydrogel obtained from water solvent was about 70 g-water/g-dry gel. Around 60-80% of the prepared starch hydrogels in this study was biologically degraded in activated sludge after 20 days, indicating their good biodegradability. [59]

Grafting of starch with other polymers was another example of improving the water absorbency. The grafting of starch backbone with acrylamide in the presence of ceric ammonium nitrate as a free radical initiator, followed by alkali saponification of the obtained graft-copolymer resulted in a hydrogel with an absorbency of 425 g water/g dry hydrogel. The increase of the starch-g-copolymer concentration has led to an improvement in the water holding capacity, where a 0.5% of the copolymer has retained almost 19% of moisture after 30 days without any further watering. In addition, the amendment of the soil with this starch-based hydrogel has decreased the soil bulk density and has improved the soil micronutrients and organic carbon. Also, there was an increase in the soil bacteria (16%) and the soil fungi count (18%). [58] Figure 12 shows the water absorbency of starch-based hydrogels.

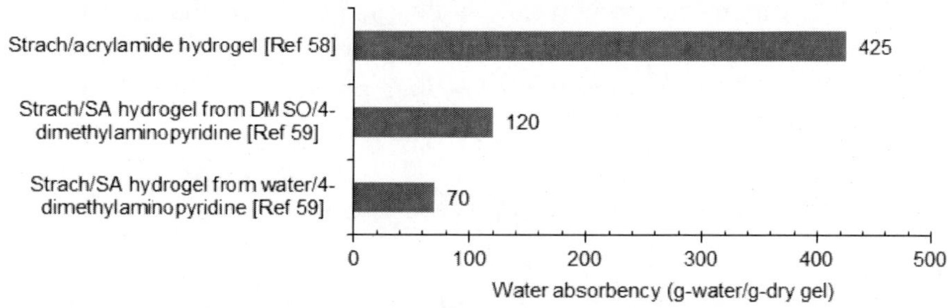

Figure 12. Water absorbency of starch-based polymers.

4.2.3. Chitin As an Absorbent Polymer

Chitin is a natural liner polymer composed of N-acetylglucosamine units (Figure 13). It is structurally similar to cellulose where the N-acetylglucosamine units form covalent β-1,4 linkages similar to the linkages between the glucose units forming cellulose. Therefore, chitin can be described as cellulose with one hydroxyl group on each monomer replaced with an acetyl amine group. The N-acetylglucosamine units of chitin allow for increased hydrogen bonding between the adjacent polymer chains, leading the chitin-polymer matrix to have an enhanced strength compared to cellulose. Chitin is of considerable interest in the development of environment friendly and biocompatible absorbent materials, because it is a non-toxic, biodegradable and biocompatible polysaccharide. In addition, it is the second most abundant polysaccharide on the earth next to cellulose.

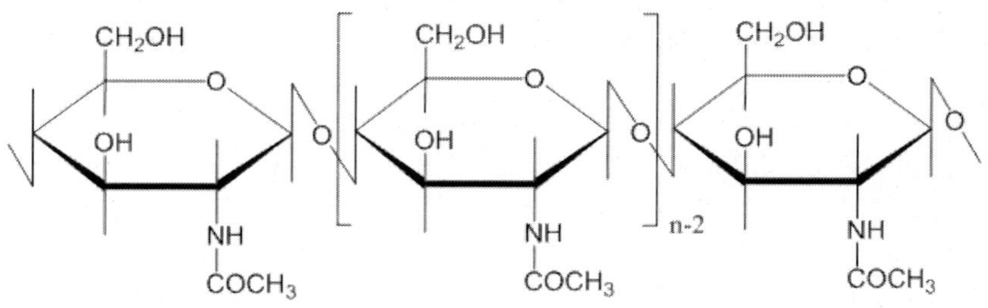

Figure 13. The chemical structure of chitin (a).

Several techniques were reported for the fabrication of chitin-superabsorbent polymers. For example, a novel superabsorbent hydrogel was obtained by a simple reaction of chitin with succinic anhydride in the presence of 4-dimethylaminopyridine as an esterification catalyst in a mixture of tetrabutylammnium fluoride and dimethylsulfoxide (TBAF/DMSO), followed by sodium hydroxide neutralization. [60] The water absorbency of the product was measured by the tea-bag method, where the hydrogel/tea-bag was immersed in water at 25° C, and then, after 2h of water treatment, the tea-bag was picked up, and the excess water was drained.

The same process was used to evaluate the absorbency after 24, 48, and 96 h. Further, the absorbency in aqueous sodium chloride solution similar to saline and seawater (concentration; 0.9% and 3.5%) was investigated. The results of this study showed that, the chitin hydrogel was obtained without any crosslinking agent when TBAF/DMSO mixture was used as a solvent, suggesting that the hydrogel was formed through the partial formation of diester between hydroxyl group of chitin and succinic anhydride. The water absorbency of this hydrogel was 330 g-water/g-dry gel, which is comparable with the conventional sodium polyacrylate. Furthermore, this chitin hydrogel exhibited a higher absorbency in aqueous sodium chloride solution than the conventional sodium polyacrylate superabsorbent hydrogel.

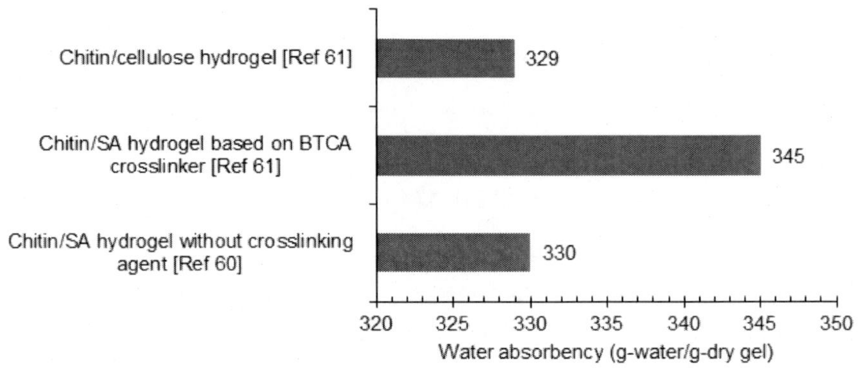

Figure 14. Water absorbency of chitin-based hydrogel.

In addition, the maximum degradation of this chitin hydrogel was 91% after 20 days, indicating its good biodegradability. In the same regard, biodegradable superabsorbent hydrogel was prepared from chitin dissolved in lithium chloride and N-methyl-2-pyrrolidinone through esterification crosslinking with 1,2,3,4-butanetetracarboxylic dianhydride (BTCA). The water absorbency of this chitin hydrogel was strongly dependent on the feeding ratio of BTCA to chitin. The hydrogel prepared at the feeding ratio of 5 showed the highest absorbency (345g water/g-dry gel). This hydrogel also exhibited a good biodegradability by chitinase with a maximum degradation of 91% within 7 days. [61] Cellulose/chitin hybrid hydrogels were also obtained from the homogeneous mixture of cellulose and chitin by a similar process to that of the preparation of the chitin hydrogel. The water absorbancy of cellulose/chitin hydrogel was 329 g-water/g-dry gel. These cellulose/chitin hybrid hydrogels were degraded by cellulase as well as chitinase. [61] Figure 14 summarizes the water absorbency of the discussed chitin-based hydrogels.

4.2.4. Alginate As an Absorbent Polymer

Alginate (Alg) is a copolymer of Beta-D-mannuronic acid and Alpha-L-guluronic acid linked together in varying proportions by 1,4-linkages (Figure 15). [62]

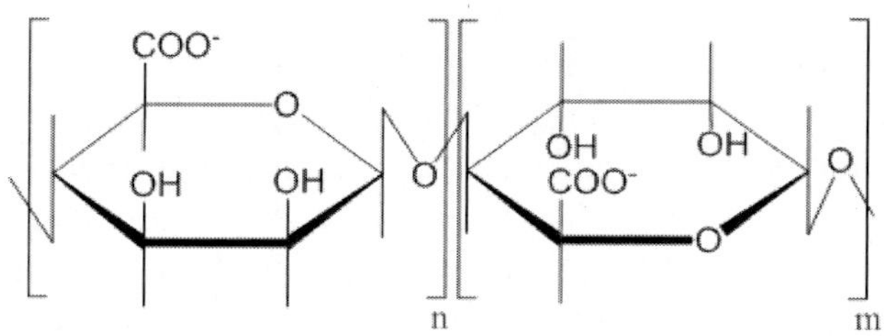

Figure 15. The chemical structure of alginate.

Alg was crosslinked with PAAm by using electron beam irradiation to form a superabsorbent hydrogel. The effect of irradiation dose on the hydrogel content and the swelling behavior were investigated in details. The increase of the irradiation dose has led to a decrease in both the hydrogel content and water absorption capacity. The water absorbency was also found to increase with the increase of Alg content in the copolymer. The maximum water absorbency was recorded to be ~600 g-water/g-dry gel (about 600 times of its dry weight) at a copolymer composition of 20/80 wt/wt of Alg/PAAm after an irradiation dose of 5 kGy. To determine the water retention, Alg/PAAm hydrogels of different weights were mixed with fixed amounts of a sandy soil. The hydrogel/soil mixtures were irrigated with the proper amount of water at 25°C. The weight loss of the mixtures against time was calculated at interval times. When the hydrogel content increased in the soil, the water retention has increased. The highest water retention was recorded when the copolymer content was 2 wt% of Alg/PAAm composition (20/80 wt/wt). Field evaluation of the Alg/PAAm hydrogel for its possible uses in agriculture as a soil conditioner was performed using bean as a model plant. It was clear that the growth and total dry weight of bean plant cultivated in the soil treated with Alg/PAAm hydrogel is greater than those of bean plant cultivated in hydrogel-free soil (control). The results also showed that the bean root can absorb large quantities of water due to the improvement of the water retention resulting from the presence of the hydrogel particles around the root system. [63]

Alg was also crosslinked with polyacrylonitrile (PAN) through an alkaline hydrolysis process of their physical mixture to prepare a superabsorbent hydrogel. For more details, during the alkaline hydrolysis process, the nitrile groups of PAN were converted to a mixture of hydrophilic carboxamide and carboxylate groups which can undergo an in situ crosslinking process with the Alg polymer. The effects of different parameters such as alkalization time, alkalization temperature, NaOH concentration, PAN/Alg weight ratio, alkaline hydrolysis temperature, alkaline hydrolysis time, post-neutralization pH and stirring speed on the hydrogel water absorbency were deeply investigated. Under the optimized conditions concluded (an alkalization time of 60 min, an alkalization temperature of 70°C, a NaOH concentration of 10 wt%, a PAN/Alg weight ratio of 2, an alkaline hydrolysis temperature of 100°C, an alkaline hydrolysis time of 90 min, a post-neutralization pH of 7, and a stirrer speed of 250 rpm), the maximum capacity of swelling in distilled water was reached 610 g-water/g-dry gel. [64]

A novel sodium Alg-*grafted*-poly(acrylic acid)/sodium humate absorbent hydrogel was prepared by graft copolymerization of sodium Alg, acrylic acid and sodium humate in aqueous solution, using N,N'-methylenebisacrylamide as a crosslinker and ammonium persulphate as an initiator. The effects of crosslinker amount, sodium Alg and sodium humate content on water absorbency of the obtained gel were studied in details. The introduction of only 10 wt% sodium humate into the sodium Alg-*grafted*-PAA system has enhanced the water absorbency of the formed hydrogel to reach 1380 g-water/g-dry gel. [65] In the same regard, A pH-sensitive semi-interpenetrating polymer network superabsorbent hydrogel composed of sodium Alg-*grafted*-poly(sodium acrylate) network and linear polyvinylpyrrolidone (PVP) was prepared by a free-radical graft polymerization in the presence of ammonium persulphate initiator and N,N'-methylenebisacrylamide crosslinker. As a result of the introduction of PVP, the surface morphologies of the prepared hydrogels were improved, leading to an enhancement in the water absorption capacity and the swelling rate of the hydrogels. The maximum water absorbency (1004 g-water/g-dry gel) was achieved at 15 wt% of PVP and 11.08 wt% of Na-Alg. [66] Figure 16 displays the water absorbency of the discussed alginate-based hydrogels.

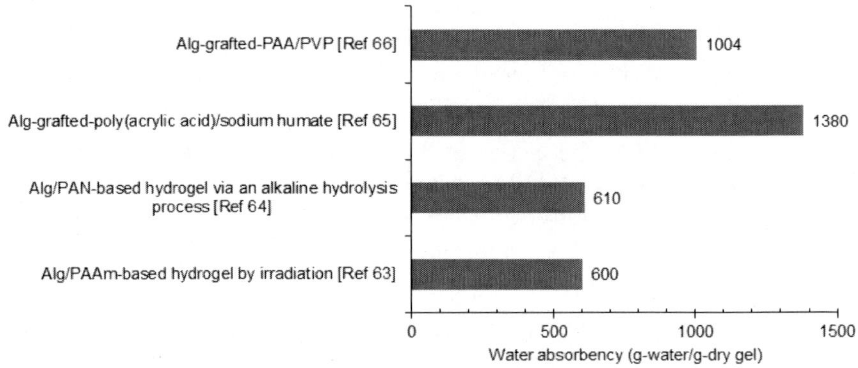

Figure 16. Water absorbency of Alginate-based hydrogels.

4.2.5. Poly(Amino Acids) As Absorbent Polymers

Biodegradable hydrogels include crosslinked poly(amino acids) such as poly(aspartic acid) (PAsp) (Figure 17a) was reported to show superabsorbent water properties.

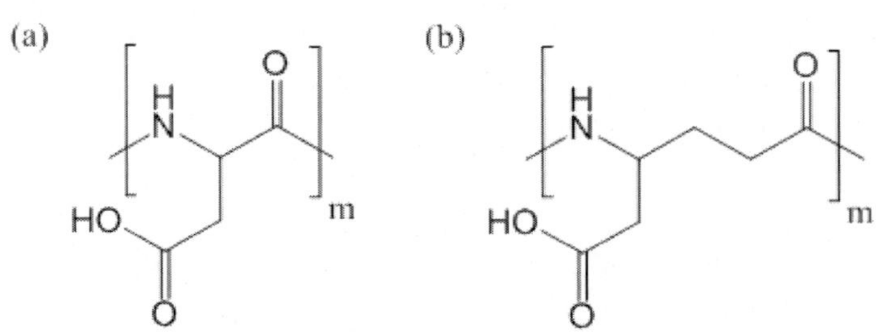

Figure 17. The chemical structure of: PAsp (a) and PGA (b).

PAsp-based hydrogels were prepared by the gamma-irradiation of PAsp produced by thermal polycondensation reaction with an acid catalyst. The effects of PAsp molecular weight, pH of the medium, concentration of PAsp in aqueous solution and the dosage of gamma-irradiation on the PAsp hydrogel preparation were investigated in details. The maximum water absorbency of this hydrogel type was recorded to be 3400 g-water/g-dry gel. The biodegradation process was checked using the activated sludge. The results showed that the PAsp hydrogel exhibited about 50% biodegradation during 28 days. [67] PAsp hydrogel was also synthesized through a chemical crosslinking process using a crosslinking agent i.e. diamine. The effects of reaction variables, such as terminal pH, reaction time and reaction temperature on the water absorbency were studied. The swelling capacity was raised by enhancing the terminal pH. The highest water absorbency was recorded to be ~420 g-water/g-dry gel at pH of 10. [68]

Poly(γ-glutamic acid) (PGA) (Figure 17b) as a poly(amino acid) was also reported to show superabsorbent water properties. PGA was crosslinked with L-lysine in aqueous solutions in the presence of 4-(4,6-dimethoxy-1,3,5-triazin-2-yl)-4-methylmorpholinium chloride (DMT-MM) or carbodiimide as crosslinking agents. The yields of the gels prepared with DMT-MM crosslinker were higher than those prepared with the water-soluble carbodiimide crosslinker. The water absorptivity of the PGA gel was affected by the amount of the crosslinker, which has an effect on the crosslinking density of the hydrogel. The water absorptivity of the PGA hydrogels crosslinked with L-lysine in the presence of DMT-MM ranged from 300 to 2100 g-water/g-dry gel, whereas the water absorptivity of the PGA hydrogels crosslinked with L-lysine in the presence of carbodiimide ranged from ~600 to ~2800 g-water/g-dry gel. The maximum water absorption was recorded at L-lysine solution pH of 5~6. [69] Figure 18 summarizes the water absorbency of PAsp-based and PGA-based hydrogels.

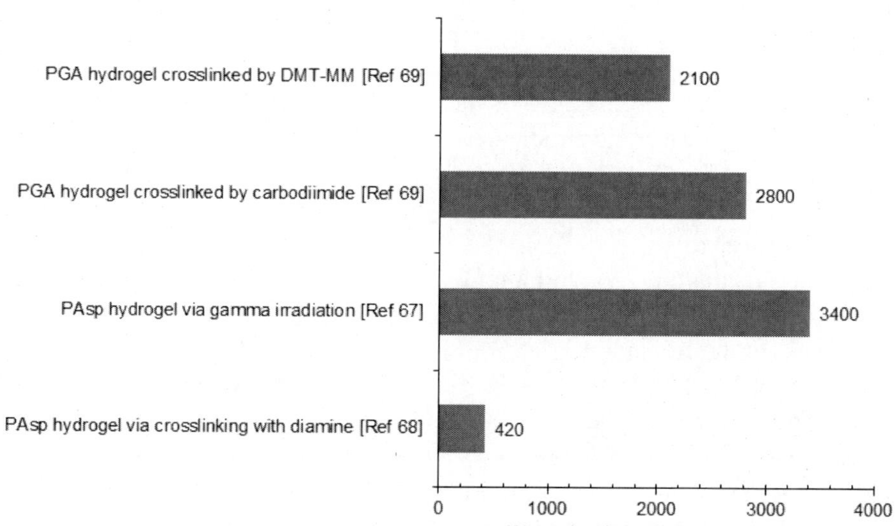

Figure 18. Water absorbency of PAsp-based and PGA-based hydrogels.

4.3. Organic-Inorganic Superabsorbent Composites

Recently, the preparations of organic-inorganic superabsorbent composites are receiving a great deal of attention because of their relative low production cost, high water absorbency, and considerable applications in agriculture and horticulture. [70-72].

4.3.1. Organic-Inorganic Superabsorbent Composites Based on Starch

A novel starch-*g*-PAAm/attapulgite superabsorbent composite was synthesized by a graft copolymerization reaction of starch, acrylamide, and attapulgite (APT) micropowder using N,N'-methylenebisacrylamide (MBA) as a crosslinker and ammonium persulphate (APS) as an initiator in an aqueous solution, followed by hydrolysis with sodium hydroxide. The initiator content, the monomers and the attapulgite ratios, and the crosslinker amount all were found to have remarkable effects on the water absorbency of the formed composite. [73] It was observed that with the increase in the content of APS initiator, the water absorbency of the composite was increased, and reached a maximum of 1317 g-water/g composite at an APS content of 1.5 wt%. A further increase in APS resulted in the decline of water absorbency of the composite. The maximum water absorbency is also reached when the composite was synthesized with a weight ratio of acrylamide to starch of 3:1 with an APT content of 10wt%. When the amount of the monomers increased, both grafting and the molecular weight of the grafted PAAm chains were increased, which resulted in an increase in the water absorbency. With further increasing of the acrylamide amount, the water absorbency was decreased. [73] This result is attributed to an increase in the homopolymer percentage of PAAm, which in turn resulted in an increase in the soluble materials at fixed crosslinking density. Furthermore, the water absorbency was decreased with the increase of the APT content. As it is well known, the inorganic clay mineral particles in the network act as an additional network points. Thus, the crosslinking density of the tarch-*g*-PAAm/APT composites increased with the increase of the APT content, which resulted in a decrease in water absorbency. The water absorbency was also decreased with the increase of MBA crosslinker content from 0.03 wt% to 0.12 wt%. When the crosslinker content was below 0.03 wt%, the composite was semi-soluble and the water absorbency was hardly to be measured. The results were due to the fact that the network of the composite cannot be formed efficiently because of the few crosslinking points when the crosslinker content is low. On the other hand, higher crosslinker content resulted in the generation of more crosslinking points, which in turn caused the formation of an additional network and then decreased the space left for water to enter the network.

Also in the same regard, a novel starch phosphate-*g*-acrylamide/attapulgite superabsorbent composite was prepared by graft-copolymerization among starch phosphate, acrylamide, and APT in aqueous solution. In this work, factors influencing water absorbency of the superabsorbent composite such as the molar ratio of NaOH, acrylamide content and the amount of starch phosphate and APT were investigated. The superabsorbent composite acquired the highest equilibrium water absorbency of 1268 g-water/g-composite (Figure 19) when the molar ratio of COO^-, COOH, and $CONH_2$ was 10:3:11, the weight ratio of acrylamide to starch phosphate was 5:1, and 10 wt% APT were incorporated. [74]

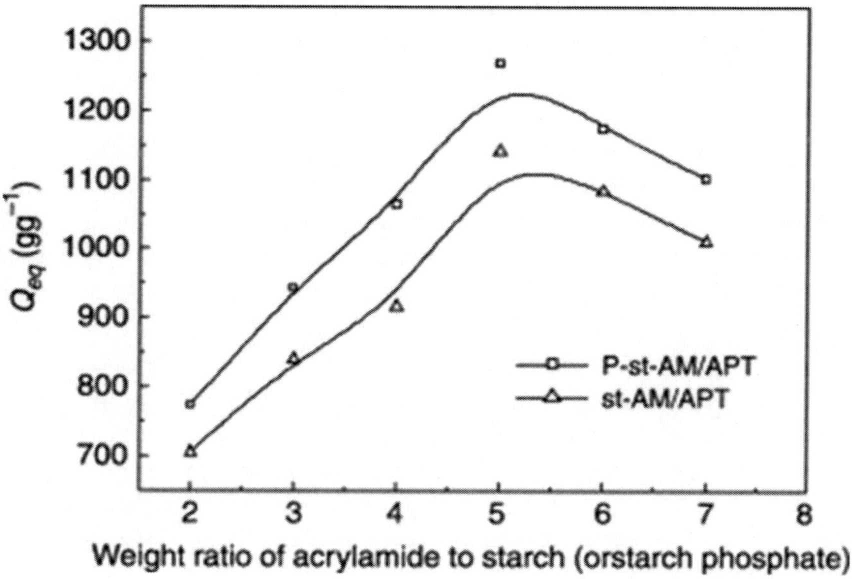

Figure 19. Equilibrium water absorbency (Q*eq*) in distilled water as a function of weight ratio of acrylamide to starch phosphate or starch for the starch-graft-acrylamide/attapulgite (st-AM/APT) and starch phosphate-graft-acrylamide/attapulgite (P-st-AM/APT) superabsorbent composite: molar ratio of NaOH to acrylamide is 0.60, APT content in the feed was 10 wt%. (Reprinted from Carbohydrate Polymers vol. 65, pp. 150–158, 2006, with permission from Elsevier publisher, License number: 3475250499169).

The effect of the clay mineral type on the composite water absorbency was also investigated. Bentonite, kaolinite and sercite clays were used to prepare clay/starch /acrylamide composites using the same conditions and compositions: 20% clay, 20% starch, 60% acrylamide monomer, 2% initiator and 0.04% crosslinker. The water absorbency of the clay-based composites were ~1600, ~2250 and ~1400 g-water/g-composite for bentonite-based composite, kaolinite-based composite and sercite-based composite, respectively. [75]

The starch-based composites were used to improve the utilization and the water-retaining of the fertilizers which are hard to be utilized by plants e.g. the phosphate rock (PHR). As a traditional fertilizer, PHR is directly used in soil. Generally, the content of available phosphorus (comprises water-soluble phosphorus and citric acid soluble phosphorus) in PHR is about 5 wt%, that means the amount of phosphorus which the plant can directly use is very small. Therefore, the water-insoluble trait of PHR limits its direct soil application. Besides, PHR is prone to deposit during the polymerizing process, which makes it difficult to directly disperse and enwrap PHR in absorbent polymers. The composite based on sulfonated starch/PAA embedding PHR (SS/PAA/PHR) has showed a good dispersion of the PHR, providing an increased content of phosphorus (P) and potassium (K) nutrients into the soil. Moreover, an excellent sustained-release behavior of the plant nutrient P and K from the SS/PAA/PHR composite was recorded. The water absorption capacity of the prepared composite was also improved to reach 498 g-water/g-composite. [76]

4.3.2. Chitosan/Acrylic Acid-Based Superabsorbent Composites

A novel chitosan-*g*-poly(acrylic acid)/montmorillonite superabsorbent nanocomposite with water absorbency of 160.1 g-water/g-composite in distilled water and 46.6 g-water/g-composite in 0.9 wt% NaCl solution was prepared by *in situ* intercalative polymerization among chitosan, acrylic acid, and montmorillonite in an aqueous solution, using MBA as a crosslinker and APS as an initiator. Chitosan intercalated into the layers of montmorillonite formed superabsorbent nanocomposites through *in situ* graft polymerization with acrylic acid (AA). The btained nanocomposites showed a high swelling ability and a pH-responsivity. Factors influencing water absorbency of the superabsorbent nanocomposite, such as weight ratio of AA to chitosan and montmorillonite content were investigated. It was evident that the montmorillonite content is an important factor influencing water absorbency of the nanocomposite.Water absorbency of the nanocomposite in distilled water and in 0.9 wt% NaCl solution was increased from 150.3 and 43.4 g-water/g-composite to 160.1 and 46.6 g-water/g-composite, respectively, as a 2 wt% of montmorillonite was introduced. The introduced montmorillonite formed a loose and porous surface which improved the water absorbency of chitosan-g-polyacrylic acid polymeric network. [77]

4.3.3. Organic-Inorganic Superabsorbent Composites Based on PAAm

A novel superabsorbent composite, PAAm/attapulgite, from acrylamide and APT, was prepared by free-radical polymerization, using MBA as a crosslinker and APS as an initiator. The effects of hydrochloric acid (HCl) concentration, acidification time, and acidification temperature while acidifying APT, and APT heat-activation on water absorbency of the superabsorbent composite in distilled water were studied in details. The water absorbency first decreased with increasing the HCl concentration while acidifying APT, and then increased with further increasing of the HCl concentration. The prolongation of acidification time was also of benefit to the increase of water absorbency. At a given HCl concentration, the water absorbency for the composite was increased with the increase of the acidification temperature. An important increase in the water absorbency was observed after the incorporation of the heat-activated APT into the polymeric network, reaching a maximum of 1964 g-water/g-composite with the APT heat-activated at 400°C. [78]

4.3.4. Organic-Inorganic Superabsorbent Composites Based on PAA

A novel poly (acrylic acid)/sodium humate superabsorbent composite was synthesized by a graft copolymerization reaction of AA on sodium humate micro powder using MBA as a crosslinker and potassium peroxydisulfate (KPS) as an initiator in an aqueous solution. The effects on water absorbency of factors such as reaction temperature, initial monomer concentration, degree of neutralization of AA, amount of crosslinker, initiator ratio, and sodium humate content were investigated. The water absorbency of the composite synthesized under optimal conditions of sodium humate content of 5.3 wt% exhibited absorption of 684 g-water/g-composite in distilled water. Furthermore, this superabsorbent composite enhanced the water retention of the soil and it also made plants grow sturdily. [79]

In the same regard, a novel superabsorbent nanocomposite was synthesized through the intercalation polymerization of partially neutralized AA and a sodium-type montmorillonite powder with MBA as a crosslinker and APS and sodium sulfite as a type of mixed redox initiators. The effects of the amounts of the sodium-type montmorillonite, crosslinker, initiator, and neutralization degree on water absorbency of the nanocomposite were deeply investigated. The results showed that AA monomer was successfully intercalated into the montmorillonite layers. The montmorillonite layers were exfoliated and basically dispersed in the composite on a nanoscale after the polymerization process. The water absorbency of the nanocomposite was much higher than that of pure poly(acrylic acid). The optimum absorbency of the nanocomposite in distilled water was 1201 g-water/g-composite. [80]

A series of novel graphene oxide (GO)/poly (acrylic acid) superabsorbent hydrogel nanocomposites were prepared via *in situ* radical solution polymerization. [81] The effects of GO content on the swelling behavior and the swelling kinetic behavior of the hydrogels were also evaluated. It was worth noting that the hydrogel containing 0.10 wt% of GO exhibited significant improvement of swelling capacity in neutral medium (1094 g-water/g-composite), and also retained relatively higher swelling capacities to a certain degree at acidic and basic solutions. The significant improvements of the swelling capacities of the GO-based nanocomposites were mainly due to the fact that the GO sheets containing many hydrophilic functional groups, such as COOH, CO, and OH groups. Besides, the GO sheets is assumed to influence the microstructure of the polymer networks because of the exceptional nanostructure character of GO, which also could influence the swelling capacities. [82]

A novel poly (acrylic acid)/mica superabsorbent composite was synthesized by a graft polymerization reaction between partially neutralized AA and ultrafine mica mineral powder. The influences of the neutralization degree of AA, as well as of the amounts of mica and crosslinker MBA on the absorbing properties were discussed in details. As observed, the water absorptivity of the composite sharply increased with increasing the crosslinker concentration from 0.07 wt% to 0.10 wt%, but it decreased when the crosslinker concentration changed from 0.10 wt% to 0.25 wt%. This phenomenon is similar to that found for starch-*g*-PAAm/clay superabsorbent composite. [83] On the other hand, the increase of the amount of mica present in the superabsorbent nanocomposite showed a decrease in the water absorbency. When the amount of mica reached 50 wt%, the water absorbency was close to 850 g-water/g-composite, which significantly reduces the preparation costs of the superabsorbents. This decrease in water absorbency with increasing amounts of mica is assumed to the fact that the ultrafine mica powder acts as an additional network point. Hence, the higher the amount of mica powder, the larger is the crosslinking density of the composite. Hence, the available network space for water molecules to enter became smaller, and thus the water absorbency gradually decreased. When AA is neutralized with NaOH, the carboxylic acid groups turned into carboxylate groups. The influence of the AA neutralization degree on the water absorbency of the composites showed that the increase in the AA neutralization degree from 60% to 65% causes a significant increase in the water absorbency. This behavior was interpreted in terms that a cooperative absorbing effect between carboxylate and carboxylic acid groups is superior to that of either group. Since in water, the carboxylate group is negatively charged. Thus, higher carboxylate concentrations will lead to repulsion among these groups, decreasing the water absorbency. On the other hand, the carboxylic acid groups are neutral. Therefore, when carboxylate and carboxylic acid groups cooperate in a suitable ratio, higher water absorbency will occur. By summarizing the results presented above, a poly(acrylic acid)/mica superabsorbent with a water absorbency higher than 1100 g-water/g-composite was synthesized having 0.10 wt% crosslinker, 1.20 wt% initiator, 10 wt% mica, a neutralization degree of 65% and a reaction temperature of 70°C. [83]

A novel poly(acrylic acid)/attapulgite superabsorbent composite was synthesized by a graft copolymerization reaction of AA on attapulgite micropowder using MBA as a crosslinker and APS as an initiator in an aqueous solution. The effects on water absorbency of factors such as reaction temperature, initial monomer concentration, degree of neutralization of AA, amount of crosslinker, initiator ratio, and attapulgite content were investigated. The water absorbencies for these superabsorbent composites in water and saline solutions in addition to the water-retention tests were also determined. The results obtained from this study showed that the water absorbency of the superabsorbent composite synthesized under optimal synthesis conditions with an attapulgite content of 10% exhibited an absorption of 1017 g-water/g-composite and 77 g-water/g-composite in distilled water and in 0.9 wt % sodium chloride solution, respectively. [84, 85] Figure 20 summarizes the water absorbency of the organic-inorganic superabsorbent composites.

In conclusion, the composite formation through the addition of the inorganic materials e.g. clays to the organic superabsorbent polymers has showed significant improvements in the swelling capacities. The improvements were related to the hydrophilic functional groups of the clays that played an important role in the enhancement of the water absorbency. In addition, the clays have influenced the microstructure of the polymer networks, leading to remarkable changes in the swelling behaviors.

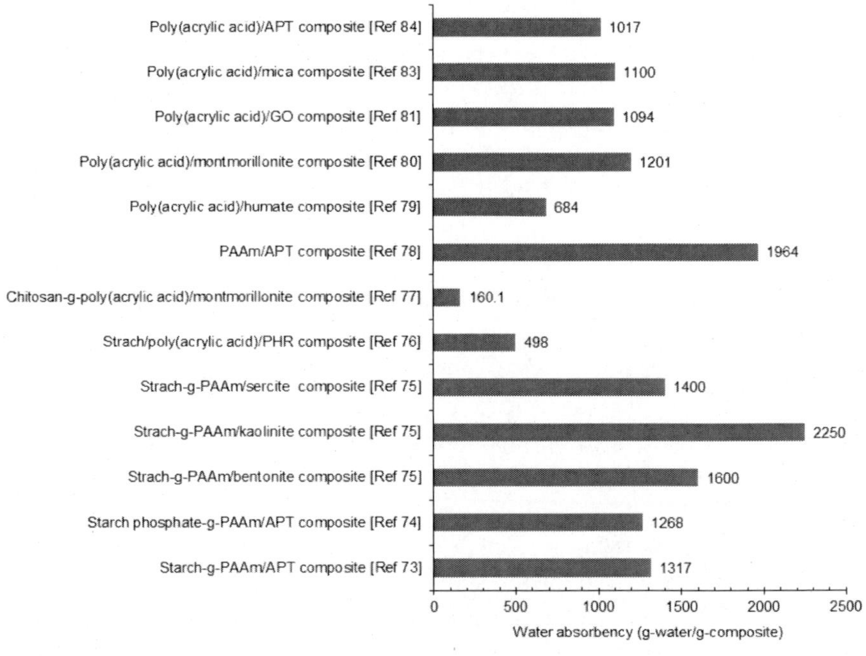

Figure 20. Water absorbency of organic-inorganic superabsorbent composites.

5. Future Impacts of Polymeric Materials in the Agricultural Area

The prospects for increasing the cultivated area are strongly connected to the development of the irrigation and drainage technology. The expected development in the agricultural production is necessarily rely on the affordability to apply new irrigation technologies, a more accurate estimation of crop water requirements, and the improvements of the irrigation construction and operation. It is also concerned with modernization procedures and methods for the proper use of surface and groundwater to minimize water use and reduce deep percolation. All these demands require enhanced research and a variety of tools such as water control and regulation equipment, remote sensing, geographic information systems, decision support systems and models, as well as field survey and evaluation techniques.

There are also some other important aspects that need to be investigated in details to determine their impact on the innovation process of water management in the irrigation sector, e.g. i-working strongly with the reuse of wastewater in irrigation. It is important to identify the pollutants that accumulate in the soil during the continuous irrigation with wastewater. Such studies will support to understand the mechanism of irrigation with wastewater. Hence, we can update the irrigation systems to match the wastewater applications. ii- the soil conditioner (absorbent polymers) must be formulated to fit the irrigation system, the soil type, and the required irrigation water dose. iii- studies elucidating the biological and the environmental impacts of the use of the absorbent polymers in irrigation systems are required to minimize their side effects and to ensure their accepted environmental levels of applications or to establish useful strategies for their use. iv- studies on the identification or the formulation of other polymeric or non-polymeric materials with specific properties of water absorption and salt separation are also required to improve the water use efficiency.

REFERENCES

[1] Laboy-Nieves E.N., Schaffner F.C., Abdelhadi A., Goosen M.F.A., *Environmental Management, Sustainable Development and Human Health.* CRC Press Taylor & Francis group. London, UK, 2009.

[2] Trimble S.W., *Encyclopedia of Water Science, Second Edition - Two Volume Set* (Print Version). Taylor & Francis, 2007.

[3] Stambouli T., Martínez-Cob A., Faci J., Howell T., Zapata N., Sprinkler evaporation losses in alfalfa during solid-set sprinkler irrigation in semiarid areas. *Irrig Sci.* 31 (2013) 1075-1089.

[4] Urrego-Pereira Y.F., Martínez-Cob A., Fernández V., Cavero J., Daytime Sprinkler Irrigation Effects on Net Photosynthesis of Maize and Alfalfa. *Agron. J.* 105 (2013) 1515-1528.

[5] Darouich H.M., Pedras C.M.G., Gonçalves J.M., Pereira L.S., Drip vs. surface irrigation: A comparison focussing on water saving and economic returns using multicriteria analysis applied to cotton. *Biosystems Engineering.* 122 (2014) 74-90.

[6] Tripathi V., Rajput T.B.S., Patel N., Performance of different filter combinations with surface and subsurface drip irrigation systems for utilizing municipal wastewater. *Irrig Sci.* 32 (2014) 379-391.

[7] Zhang Z., Hu H., Tian F., Hu H., Yao X., Zhong R., Soil salt distribution under mulched drip irrigation in an arid area of northwestern China. *Journal of Arid Environments.* 104 (2014) 23-33.

[8] Rowe D.B., Kolp M.R., Greer S.E., Getter K.L., Comparison of irrigation efficiency and plant health of overhead, drip, and sub-irrigation for extensive green roofs. *Ecological Engineering*. 64 (2014) 306-313.

[9] Osman K.T., Physical Properties of Forest Soils, *Forest Soils*. Springer International Publishing, 2013.

[10] Gil M., Rodríguez-Sinobas L., Juana L., Sánchez R., Losada A., Emitter discharge variability of subsurface drip irrigation in uniform soils: effect on water-application uniformity. *Irrig Sci*. 26 (2008) 451-458.

[11] Bhattarai S.P., Midmore D.J., Pendergast L., Yield, water-use efficiencies and root distribution of soybean, chickpea and pumpkin under different subsurface drip irrigation depths and oxygation treatments in vertisols. *Irrig Sci.* 26 (2008) 439-450.

[12] Tl T., Pang H.-c., Li Y.-y., The Potential Contribution of Subsurface Drip Irrigation to Water-Saving Agriculture in the Western USA. *Agricultural Sciences in China*. 8 (2009) 850-854.

[13] Cook F.J., Thorburn P.J., Bristow K.L., Cote C.M., Infiltration from surface and buried point sources: The Average wetting water content. *Water Resources Research.* 39 (2003) 1364.

[14] Shaviv A., Sinai G., Application of Conditioner Solution by Subsurface Emitters for Stabilizing the Surrounding Soil. *Journal of Irrigation and Drainage Engineering*. 130 (2004) 485-490.

[15] Ben-Hur M., Using synthetic polymers as soil conditioners to control runoff and soil loss in arid and semi-arid regions—a review. *Soil Research.* 44 (2006) 191-204.

[16] Zumsteg R., Plötze M., Puzrin A., Effect of Soil Conditioners on the Pressure and Rate-Dependent Shear Strength of Different Clays. *Journal of Geotechnical and Geoenvironmental Engineering*. 138 (2012) 1138-1146.

[17] Dragan E.S., Design and applications of interpenetrating polymer network hydrogels. A review. *Chemical Engineering Journal*. 243 (2014) 572-590.

[18] Hedrick R.M., Mowry D.T., Effect of Synthetic Polyelectrolytes on Aggregation, Aeration, and Water Relationships of Soil. *Soil Science*. 73 (1952) 427-442.

[19] Douglas G.H., Sathiskumar D., Larry C.W., Karen K.L., Carol M., Debra A.I., *Biodegradable Agricultural Mulches Derived from Biopolymers, Degradable Polymers and Materials: Principles and Practice* (2nd Edition). American Chemical Society, 2012.

[20] Akelah A., Polymers in Plantation and Plants Protection, *Functionalized Polymeric Materials in Agriculture and the Food Industry*. Springer US, 2013.

[21] Orts W.J., Sojka R.E., Glenn G.M., Biopolymer additives to reduce erosion-induced soil losses during irrigation. *Industrial Crops and Products.* 11 (2000) 19-29.

[22] Hafez I.H., Zein-El-Abedin T.K., Mourad A.Y., Abd-El-Hady A.M., Improving the wetting front distribution in sandy soil using soil conditioners under subsurface drip irrigation *Master thesis*. (2008) 1-139.

[23] Xindong W., Xuefeng Y., Yumei L., Youke W., Research on the Water-saving and Yield-increasing Effect of Polyacrylamide. *Procedia Environmental Sciences*. 11, Part B (2011) 573-580.

[24] Akhter J., Mahmood K., Malik K.A., Mardan A., Ahmad M., Iqbal M.M., Effects of hydrogel amendment on water storage of sandy loam and loam soils and seedling growth of barley, wheat and chickpea. *Plant, Soil and Environment* 50 (2004) 463-469.

[25] El-Hady O.A., Abd El-Kader A.A., Shafi A.M., Physico-bio-chemical Properties of Sandy Soil Conditioned with Acrylamide Hydrogels after Cucumber Plantation. *Australian Journal of Basic & Applied Sciences*. 3 (2009) 3145.

[26] Sivapalan S., Benefits of treating a sandy soil with a crosslinked-type polyacrylamide. *Australian Journal of Experimental Agriculture*. 46 (2006) 579-584.

[27] Liu Z.S., Rempel G.L., Preparation of superabsorbent polymers by crosslinking acrylic acid and acrylamide copolymers. *Journal of Applied Polymer Science.* 64 (1997) 1345-1353.

[28] Karadağ E., Saraydin D., Güven O., Water absorbency studies of γ-radiation crosslinked poly(acrylamide-co-2,3-dihydroxybutanedioic acid) hydrogels. *Nuclear Instruments and Methods in Physics Research Section B: Beam Interactions with Materials and Atoms*. 225 (2004) 489-496.

[29] Lu Y., Jing R., Kong Q.-m., Zhu P.-x., Solid state grafting copolymerization of acrylamide onto poly(vinyl alcohol) initiated by redox system. *Journal of Applied Polymer Science*. 131 (2014) n/a-n/a.

[30] TUDORACHI N., LIPSA R., Copolymers based on poly(vinyl alcohol) and acrylamide. *Journal of Optoelectronics and Advanced Materials* 8(2006) 659-662.

[31] Nadler A., Letey J., Adsorption Isotherms of Polyanions on Soils Using Tritium Labeled Compounds. *Soil Sci. Soc. Am. J.* 53 (1989) 1375-1378.

[32] Malik M., Letey J., Adsorption of Polyacrylamide and Polysaccharide Polymers on Soil Materials. *Soil Sci. Soc. Am. J.* 55 (1991) 380-383.

[33] Aly S.M., Letey J., Polymer and Water Quality Effects on Flocculation of Montmorillonite. *Soil Sci. Soc. Am.* J. 52 (1988) 1453-1458.

[34] Nadler A., Malik M., Letey J., Desorption of polyacrylamide and polysaccharide polymers from soil materials. *Soil Technology*. 5 (1992) 91-95.

[35] Lu J.H., Wu L., Letey J., Effects of Soil and Water Properties on Anionic Polyacrylamide Sorption. *Soil Sci. Soc. Am. J.* 66 (2002) 578-584.

[36] Loiseau J., Doërr N., Suau J.M., Egraz J.B., Llauro M.F., Ladavière C., Claverie J., Synthesis and Characterization of Poly(acrylic acid) Produced by RAFT Polymerization. Application as a Very Efficient Dispersant of CaCO3, Kaolin, and TiO2. *Macromolecules*. 36 (2003) 3066-3077.

[37] Guiwei Q., De Varennes A., Cunha-Queda C., Remediation of a mine soil with insoluble polyacrylate polymers enhances soil quality and plant growth. *Soil Use and Management*. 24 (2008) 350-356.

[38] Achtenhagen J., Kreuzig R., Laboratory tests on the impact of superabsorbent polymers on transformation and sorption of xenobiotics in soil taking 14C-imazalil as an example. *Science of The Total Environment*. 409 (2011) 5454-5458.

[39] Durovic N., Pivic R., Pocuca V., Effects of the application of a hydrogel in different soils. *Agriculture & Forestry*. 53 (2012) 25-34.

[40] Orikiriza L.J.B., Agaba H., Tweheyo M., Eilu G., Kabasa J.D., Hüttermann A., Amending Soils with Hydrogels Increases the Biomass of Nine Tree Species under Non-water Stress Conditions. *CLEAN* – Soil, *Air, Water.* 37 (2009) 615-620.

[41] Lawrence J.B.O., Hillary A., Gerald E., John D.K., Martin W., Aloys H., Effects of Hydrogels on Tree Seedling Performance in Temperate Soils before and after Water Stress. *Earth & Environmental Sciences*. 4 (2013) 713-721.

[42] Zhuang W., Li L., Liu C., Effects of sodium polyacrylate on water retention and infiltration capacity of a sandy soil. *SpringerPlus*. 2 (2013) 1-7.
[43] Elliott J.E., Macdonald M., Nie J., Bowman C.N., Structure and swelling of poly(acrylic acid) hydrogels: effect of pH, ionic strength, and dilution on the crosslinked polymer structure. *Polymer*. 45 (2004) 1503-1510.
[44] Francis S., Kumar M., Varshney L., Radiation synthesis of superabsorbent poly(acrylic acid)–carrageenan hydrogels. *Radiation Physics and Chemistry*. 69 (2004) 481-486.
[45] Wei Q., Fast-swelling porous starch-g-poly(acrylic acid) superabsorbents. *Iran Polym J*. 23 (2014) 637-643.
[46] Feng D., Bai B., Ding C., Wang H., Suo Y., Synthesis and Swelling Behaviors of Yeast-g-Poly(acrylic acid) Superabsorbent Co-polymer. *Industrial & Engineering Chemistry Research*. 53 (2014) 12760-12769.
[47] Zohuriaan-Mehr M.J., Kabiri K., Superabsorbent Polymer Materials: A Review. *Iran Polym J*. 17 (2008) 451-477.
[48] Esposito F., Del Nobile M.A., Mensitieri G., Nicolais L., Water sorption in cellulose-based hydrogels. *Journal of Applied Polymer Science*. 60 (1996) 2403-2407.
[49] Pan S., Ragauskas A.J., Preparation of superabsorbent cellulosic hydrogels. *Carbohydrate Polymers*. 87 (2012) 1410-1418.
[50] Yang L., Zhang H.-Y., Yang Q., Lu D.-n., Bacterial cellulose–poly(vinyl alcohol) nanocomposite hydrogels prepared by chemical crosslinking. *Journal of Applied Polymer Science*. 126 (2012) E245-E251.
[51] Xu J.B., Bartley J.P., Johnson R.A., Preparation and characterization of alginate–carrageenan hydrogel films crosslinked using a water-soluble carbodiimide (WSC). *Journal of Membrane Science*. 218 (2003) 131-146.
[52] Sannino A., Pappadà S., Madaghiele M., Maffezzoli A., Ambrosio L., Nicolais L., Crosslinking of cellulose derivatives and hyaluronic acid with water-soluble carbodiimide. *Polymer*. 46 (2005) 11206-11212.
[53] Demitri C., Del Sole R., Scalera F., Sannino A., Vasapollo G., Maffezzoli A., Ambrosio L., Nicolais L., Novel superabsorbent cellulose-based hydrogels crosslinked with citric acid. *Journal of Applied Polymer Science*. 110 (2008) 2453-2460.
[54] Yoshimura T., Matsuo K., Fujioka R., Novel biodegradable superabsorbent hydrogels derived from cotton cellulose and succinic anhydride: Synthesis and characterization. *Journal of Applied Polymer Science*. 99 (2006) 3251-3256.
[55] Fei B., Wach R.A., Mitomo H., Yoshii F., Kume T., Hydrogel of biodegradable cellulose derivatives. I. Radiation-induced crosslinking of CMC. *Journal of Applied Polymer Science*. 78 (2000) 278-283.
[56] Wach R.A., Mitomo H., Yoshii F., Kume T., Hydrogel of biodegradable cellulose derivatives. II. Effect of some factors on radiation-induced crosslinking of CMC. *Journal of Applied Polymer Science*. 81 (2001) 3030-3037.
[57] Andry H., Yamamoto T., Irie T., Moritani S., Inoue M., Fujiyama H., Water retention, hydraulic conductivity of hydrophilic polymers in sandy soil as affected by temperature and water quality. *Journal of Hydrology*. 373 (2009) 177-183.
[58] Parvathy P.C., Jyothi A.N., John K.S., Sreekumar J., Cassava Starch Based Superabsorbent Polymer as Soil Conditioner: Impact on Soil Physico-Chemical and Biological Properties and Plant Growth. *CLEAN – Soil, Air, Water*. (2014) n/a-n/a.

[59] Yoshimura T., Yoshimura R., Seki C., Fujioka R., Synthesis and characterization of biodegradable hydrogels based on starch and succinic anhydride. *Carbohydrate Polymers*. 64 (2006) 345-349.
[60] Yoshimura T., Uchikoshi I., Yoshiura Y., Fujioka R., Synthesis and characterization of novel biodegradable superabsorbent hydrogels based on chitin and succinic anhydride. *Carbohydrate Polymers*. 61 (2005) 322-326.
[61] Kono H., Zakimi M., Preparation, water absorbency, and enzyme degradability of novel chitin- and cellulose/chitin-based superabsorbent hydrogels. *Journal of Applied Polymer Science*. 128 (2013) 572-581.
[62] Remminghorst U., Rehm B.A., Bacterial alginates: from biosynthesis to applications. Biotechnol Lett. 28 (2006) 1701-1712.
[63] Abd El-Rehim H.A., Characterization and possible agricultural application of polyacrylamide/sodium alginate crosslinked hydrogels prepared by ionizing radiation. *Journal of Applied Polymer Science*. 101 (2006) 3572-3580.
[64] Marandi G.B., Sharifnia N., Hosseinzadeh H., Synthesis of an alginate–poly(sodium acrylate-co-acrylamide) superabsorbent hydrogel with low salt sensitivity and high pH sensitivity. *Journal of Applied Polymer Science*. 101 (2006) 2927-2937.
[65] Hua S., Wang A., Synthesis, characterization and swelling behaviors of sodium alginate-g-poly(acrylic acid)/sodium humate superabsorbent. *Carbohydrate Polymers*. 75 (2009) 79-84.
[66] Wang W., Wang A., Synthesis and swelling properties of pH-sensitive semi-IPN superabsorbent hydrogels based on sodium alginate-g-poly(sodium acrylate) and polyvinylpyrrolidone. *Carbohydrate Polymers*. 80 (2010) 1028-1036.
[67] Tomida M., Yabe M., Arakawa Y., Kunioka M., Preparation conditions and properties of biodegradable hydrogels prepared by γ-irradiation of poly(aspartic acid)s synthesized by thermal polycondensation. *Polymer*. 38 (1997) 2791-2795.
[68] Zhao Y., Su H., Fang L., Tan T., Superabsorbent hydrogels from poly(aspartic acid) with salt-, temperature- and pH-responsiveness properties. *Polymer*. 46 (2005) 5368-5376.
[69] Murakami S., Aoki N., Matsumura S., Bio-based biodegradable hydrogels prepared by crosslinking of microbial poly([gamma]-glutamic acid) with L-lysine in aqueous solution. *Polym J*. 43 (2011) 809-809.
[70] Bo Z., Renkuan L., Yunkai L., Tao G., Peiling Y., Ji F., Weimin X., Zhichao Z., Water-absorption characteristics of organic–inorganic composite superabsorbent polymers and its effect on summer maize root growth. *Journal of Applied Polymer Science*. 126 (2012) 423-435.
[71] Feng E., Ma G., Wu Y., Wang H., Lei Z., Preparation and properties of organic–inorganic composite superabsorbent based on xanthan gum and loess. *Carbohydrate Polymers*. 111 (2014) 463-468.
[72] Zhang S., Guan Y., Fu G.-Q., Chen B.-Y., Peng F., Yao C.-L., Sun R.-C., Organic/Inorganic Superabsorbent Hydrogels Based on Xylan and Montmorillonite. *Journal of Nanomaterials*. 2014 (2014) 11.
[73] Li A., Liu R., Wang A., Preparation of starch-graft-poly(acrylamide)/attapulgite superabsorbent composite. *Journal of Applied Polymer Science*. 98 (2005) 1351-1357.

[74] Zhang J., Li A., Wang A., Study on superabsorbent composite. VI. Preparation, characterization and swelling behaviors of starch phosphate-graft-acrylamide /attapulgite superabsorbent composite. *Carbohydrate Polymers.* 65 (2006) 150-158.
[75] Wu J., Lin J., Zhou M., Wei C., Synthesis and properties of starch-graft-polyacrylamide/clay superabsorbent composite. *Macromolecular Rapid Communications.* 21 (2000) 1032-1034.
[76] Zhong K., Lin Z.-T., Zheng X.-L., Jiang G.-B., Fang Y.-S., Mao X.-Y., Liao Z.-W., Starch derivative-based superabsorbent with integration of water-retaining and controlled-release fertilizers. *Carbohydrate Polymers.* 92 (2013) 1367-1376.
[77] Zhang J., Wang L., Wang A., Preparation and Properties of Chitosan-g-poly(acrylic acid)/Montmorillonite Superabsorbent Nanocomposite via in Situ Intercalative Polymerization. *Industrial & Engineering Chemistry Research.* 46 (2007) 2497-2502.
[78] Wang W., Zhang J., Chen H., Wang A., Study on superabsorbent composite. VIII. Effects of acid- and heat-activated attapulgite on water absorbency of polyacrylamide /attapulgite. *Journal of Applied Polymer Science.* 103 (2007) 2419-2424.
[79] Mo C., Shu-quan Z., Hua-Min L., Zhan-bin H., Shu-qin L., Synthesis of poly(acrylic acid)/sodium humate superabsorbent composite for agricultural use. *Journal of Applied Polymer Science.* 102 (2006) 5137-5143.
[80] Liu P.-S., Li L., Zhou N.-L., Zhang J., Wei S.-H., Shen J., Synthesis and properties of a poly(acrylic acid)/montmorillonite superabsorbent nanocomposite. *Journal of Applied Polymer Science.* 102 (2006) 5725-5730.
[81] Zhu Z.-Q., Sun H.-X., Qin X.-J., Jiang L., Pei C.-J., Wang L., Zeng Y.-Q., Wen S.-H., La P.-Q., Li A., Deng W.-Q., Preparation of poly(acrylic acid)-graphite oxide superabsorbent nanocomposites. *Journal of Materials Chemistry.* 22 (2012) 4811-4817.
[82] Huang Y., Zeng M., Ren J., Wang J., Fan L., Xu Q., Preparation and swelling properties of graphene oxide/poly(acrylic acid-co-acrylamide) super-absorbent hydrogel nanocomposites. *Colloids and Surfaces A: Physicochemical and Engineering Aspects.* 401 (2012) 97-106.
[83] Lin J., Wu J., Yang Z., Pu M., Synthesis and Properties of Poly(acrylic acid)/Mica Superabsorbent Nanocomposite. *Macromolecular Rapid Communications.* 22 (2001) 422-424.
[84] Li A., Wang A., Chen J., Studies on poly(acrylic acid)/attapulgite superabsorbent composite. I. Synthesis and characterization. *Journal of Applied Polymer Science.* 92 (2004) 1596-1603.
[85] Li A., Wang A., Chen J., Studies on poly(acrylic acid)/attapulgite superabsorbent composites. II. Swelling behaviors of superabsorbent composites in saline solutions and hydrophilic solvent–water mixtures. *Journal of Applied Polymer Science.* 94 (2004) 1869-1876.

In: Drip Irrigation
Editor: Alfred H. Steele
ISBN: 978-1-63463-737-4

Chapter 4

DESIGN OF CLOGGING-FREE UNIFORM EMISSION DRIP IRRIGATION SYSTEM USING MICROTUBES

Muhammed A. Bhuiyan* and Amir Keshtgar
School of Civil, Environmental and Chemical Engineering, RMIT University, Melbourne, Australia

ABSTRACT

Drip irrigation is a system in which water is supplied with a given flow rate directly to plant roots to meet its water requirement. Due to water scarcity worldwide drip irrigation practice is gaining momentum because of its better control on water volume and management strategy with time and space. The system is compatible for a wide range of crop variety, soil type, climate and landscape despite of few inherent constrains.

Since long-back clogging and emission non-uniformity have been the major obstacles in the development of drip irrigation using various emitters. Clogging would be a serious problem in areas with brackish water where the problems of precipitation of calcium carbonate, organic materials and suspended sands are severe. Installation of filter equipment to the system could not eliminate the problem entirely and thus irrigators have to use different acidic solutions to remove precipitation, which have adverse influence on soil and crops. On the other hand to obtain the best emission uniformity (EU) in plain/uneven lands the pressure regulators and pressure compensating emitters have been in use for a long time. However, pressure compensating emitters tend to be more complex and expensive than non-compensating emitters.

This chapter suggests the possibility of utilizing small diameter pipes approximately 2 to 4mm called microtubes, for enhancing the emission uniformity of the drip system and reducing the difficulties encountered by those emitters due to clogging and blockage. Microtubes have many advantages compared to traditional types of emitters in terms of cost and practical applications. As these pipes are made of flexible materials can be adjusted in shape and length easily. By adjusting the microtube lengths according to energy head developed throughout a network, an equal emission can be delivered to any landscape. Here microtubes act as emitters along the lengths of laterals, where laterals are emerged from manifold to form a subunit of the irrigation system.

---

* Corresponding author: Muhammed.bhuiyan@rmit.edu.au.

The variation of the microtube lengths along laterals are done for compensating the energy heads above a threshold value in a given manifold (or subunit). This threshold value has been set at the very last microtube of the end-lateral, which is equal to the frictional and other minor head loses calculated for a minimum length microtube to reach the plant root. The lengths of the other microtubes would consistently increase according to excess heads above the threshold head would have generated. Hence these laterals can also be imagined as larger emitters along the manifold of an irrigation subunit. These larger emitters along the manifold would have a characteristic pressure-discharge relationship for a hydraulically calculated set of microtube lengths emitting equal discharges along the end-lateral. At this stage there would have two options for the design of the succeeding laterals. In one option, the length of all the microtubes in the succeeding laterals will have the same set of lengths as has been calculated for the end-lateral. This configuration of the microtube setup will deliver to some extent a variable discharge; nonetheless the designer can adopt an EU value (say EU $\geq$ 90%) to allow the maximum variation by restricting the total number of laterals and thereby the length of each manifold. In other option, the designer will vary all the microtube lengths in all the successive laterals so as to deliver equal flows and thereby to achieve full emission uniformity (EU).

When the required discharge and diameters of microtube, lateral and manifold and some ground conditions are given, the length of the microtubes, the pressure heads, EU (in percentage) and the best subunit dimensions can be obtained by using this simple algorithm. The results of this study showed that larger sized microtubes would have higher variation of length, while applying larger flow rates could have decreased the microtube length variation along the laterals.

1. INTRODUCTION

1.1. Drip Irrigation

Drip or trickle also called micro or localized irrigation is a pressurized system to irrigate the crops and orchards, consists of an extensive network of pipes usually of small diameters that deliver water directly to the soil near the plant. The system usually possesses fertilizer injection system, supplying plants with needed nutrients. In drip irrigation the objective is to provide each plant with a continuous readily available supply of soil moisture, which is sufficient to meet transpiration demands (Keller and Karmeli, 1974). A filter is used to remove suspended materials, organic matter, sand and clay to reduce blockage of the emitters. Along with pumping station, control valves are installed to provide required pressure heads to the system (Hensen et al., 1980; Bralts and Wu, 1979b).

The system contains emitters, laterals, manifold and mainline, which supplies water from the source to plant root zone. The mainline delivers water to manifold and manifold delivers water to laterals. The emitters, which are attached to laterals, distribute water to plant root zones. Laterals are normally one sized pipes, made of polyethylene (PE) with diameters 10 to 16 mm in range, providing better flushing, easy installation and maintenance characteristics (Hensen et al., 1980, Perold, 1977). Manifold and mainline are either in medium density polyethylene (PE) or rigid PVC with diameters 20 to 100 mm in range (Wu and Gitlin, 1977b). Emitters are plant's point sources of water and designed to provide small and largely equal amount of discharges of plant requirement. Emitters are of many kinds, such as, orifices, nozzles, porous pipes, microtubes, etc to dissipate the pressure in the pipe distribution networks by means of a narrow nozzle or long flow path, and thereby decreasing the water pressure to allow discharges of only a few litters per hour (Vermeiren and Jobling, 1980; Bralts and Wu, 1979). Then water is distributed by its normal movement through the soil profile.

1.2. Problems of Drip Irrigation

Clogging of the small conduits in the emitters is the most serious problem of drip irrigation system. Sand and clay particles, debris, chemical precipitants and organic growth can block flows through emitters. The clogging occurs gradually and reduces the flow rates of emitters and causing poor water distribution (Vermeiren and Jobling, 1980). This problem will lead to poor productivity in farms or orchards because some parts of the land obtain more water while other parts do not meet the minimum requirement of plants due to evaporation. Clogging occurs mainly due to passage of water through the very fine pores of the emitters. Thus, emitters with smaller passages are more susceptible to blockage. The delivered water contains suspended particles, salts and dissolved fertilizers which create severe problems that require tedious efforts and skilled man-power to resolve. Using acid injection and/or replacing new emitters are the main common solutions for clogging problem which both are time consuming and impose huge running cost to the system (Gilbert et al., 1981). Moreover, uneven flow rate due to pressure head loss is another major problem of this system. Drip irrigation systems do not apply water with perfect uniformity along the crop rows. Some of the variability is caused by manufacturing imperfection in the emitters, but the major problem crops up from the stance of the system design, in terms of the frictional loss in the direction of flow through the lateral pipe or tubing where emitters are attached (Myers and Bucks, 1972).

Drip irrigation usually operates under low pressure (less than 100 kPa). Pressure distribution inside a lateral or manifold will be greatly affected by the friction and slope of the pipe laying. This variation of pressure will change the discharge of emitters along the line. However, ideal drip irrigation will be one which can irrigate uniformly that each emitter delivers equal discharge as required by the plant per irrigation (Wu and Gitlin, 1973). Several manufacturers have designed emitters to reduce discharge variations caused by friction-induced pressure changes in the lateral pipes. These emitters are complex and susceptible to clogging to impose huge initial and operational costs to the system. In another instance, a graphical procedure was developed by Wu and Gitlin (1973) using simple emitters of different diameters along the lateral pipe to compensate its frictional losses, which in practice is very hard to install.

In terms of economical consideration, drip irrigation requires high capital cost as well as maintenance. Studies showed that due to big initial and running costs, the system has limited application in small plots of the developing countries (Singh et al., 2009).

1.3. Advantages of Drip Irrigation

Drip irrigation offers unique agronomical and economical advantages for the efficient use of water. It is considered as the most water saving method of irrigation, thus very ideal for use in areas with limited water resources. It is also beneficial in temperate areas where other surface irrigations such as flood and furrow irrigations or pressurized systems such as sprinkler are subject to big water loss due to evaporation. Since the system's emission can be controlled by time management or using different types of emitters, the problem of deep percolation and loss in the soils can be reduced and saves water in root zone and subsequently application efficiency could be achieved. This system can be applied very efficiently to small trees and widely spaced plants such as tomatoes, citrus and grapes (Benami and Ofen, 1984; Wu and Gitlin, 1977b). In arid regions with good management the ratio of transpired to applied water is usually at least 0.9 (Hensen et al., 1980). Water application efficiencies approach 100 percent and water savings of 30 to 50 percent over other irrigation methods are obtained for crops and conditions favouring drip irrigation (Hensen et al., 1980). Insect, disease and fungus problems are reduced by minimizing the wetting of the soil surface. Fewer weeds, less soil crusting, reduced cultivation and thus less soil compaction interference with harvesting are other benefits of drip irrigation (Hensen et al., 1980).

The possibility of applying the fertilizer and pesticide by injection into irrigation water and deliver to plant root zone can help to decline the deep percolation of this important input and increase their efficiency. Use of saline water is not recommended in sprinkler systems, which cause leaf burn. However, saline water can be applied in drip irrigation systems. Saline water should be applied with caution because salts may cause emitter clogging and require frequent soil leaching to prevent salt accumulation in the soil (Vermeiren and Jobling, 1980).

1.4. Objectives of the Study

Microtubes also called spaghetti tubes are small bore polyethylene tubes, in the range of 1 to 4 mm in internal diameters, are used as emitters in drip irrigation system. These small-bore tubes can be used as pressure compensating emitters in drip irrigation system. Utilizing these flexible tubes as an alternative to modern dripping emitters will reduce the risk of clogging significantly as they have simpler passages than those emitters. Microtubes have been successfully used as emitters in drip irrigation system with the benefit of having equal flow rates at all outlets along the laterals without significant problem of clogging (Vermeiren and Jobling, 1980). The varied total heads along the laterals can be compensated by using a set of varied microtube lengths to provide requisite frictional losses and thereby to deliver equal flow rates to plants. Otherwise, laterals would have discharged varying flow rates according to its varying inlet total heads. These inlet heads, indeed vary due to frictional losses in the succeeding manifold reaches for its ensuing discharges.

When each microtube works as an emitter and discharges equally (emission uniformity, EU = 100%) at different points along a particular lateral, the lateral at that time can also be imagined as an emitter along the given manifold (Keller and Karmeli, 1974). Thus on the other hand, if we want to standardize a set of varied microtube lengths for an inlet head of a given lateral, it would deliver proportionately varying discharges according to the inlet heads of the successive laterals. Because this standardized set of microtube lengths are derived based on an inlet head of a given lateral to deliver equal discharges, the given lateral would have a characteristic pressure-discharge relationship, which can subsequently be applied for successive laterals with varying inlet heads to produce consequential discharges. These discharges as would vary along the manifold line, a limiting emission uniformity (EU) set by the designer, would dictate the resulting length of the manifold with the number of laterals to decide about the dimensions of the irrigation subunit.

Computes codes are written to calculate the microtube lengths and total heads along the laterals within a subunit for a given discharge or discharges set through EU. These discharges are projected according to the demand of the plants and the limiting percent of EU. The diameters of the microtube and lateral, ground slope, and spaces between plants are used to compute the total heads of the system and the length of manifold required to fulfil the considered EU.

In one code the laterals are imagined as large emitters along the manifold and the set of microtube lengths are proposed to be same for all the laterals, thus emission uniformity would dictate the subunit dimensions. In the other code, for a given subunit, varying microtube lengths along each lateral are calculated to produce equal discharge throughout the system to achieve 100 percent EU. Therefore, the specific objectives of this research are as follows:

1. To develop computer programs, one is for pre-defined EU (say, EU = 90%) set by the designer and the other one is for a Full EU system (EU = 100%).
2. Using the above computer programs for several typical scenarios, design tables and graphs would be developed to determine appropriate microtube and manifold lengths.

2. Scope of Study

A suit of computer codes have been written using Visual Basic as a graphic user interface (GUI) program which can easily be run by ordinary home computers. In order to achieve the best accuracy using the program, the designer should consider some constraints that will be discussed later. The programs that have been developed to simulate the system are as follow:

i. Determination of microtube lengths and inlet pressure head of the end-lateral – in this code the microtube and lateral diameters and microtube spacing and number should be known.
ii. Determination of length and inlet pressure head of the manifold assuming each lateral in the manifold works as a large emitter – in this part the flow rates of all successive laterals have been computed for a given EU using the results from first code and the corresponding pressure-discharge relationship. This code has been developed based on the practice of convenience that the entire system applies the same set of microtube lengths as obtained in end-lateral.
iii. Design an ideal drip system with 100 percent EU using varied microtube length all over the system – this code also works based on the results of first code to initiate and create new series of microtube lengths for each lateral separately. Design tables and graphs have been generated for direct calculation of microtube lengths.

2.1. Microtubes and Emitters

During recent years, numerous drip irrigation emitters with varying characteristics have become available in the market. To have the best emission uniformity and minimum flow rate fluctuation due to pressure distribution, some of the emitters have been designed as 'pressure compensating' emitter. Some of them are self cleaning or 'flushing' to reduce the clogging but others can be clogged easily and require sophisticated water filtration. Solomon (1979) and Keller and Karmeli (1974) among others have listed the desired qualities of those emitters in drip irrigation systems.

Drip or trickle irrigation researchers have chosen two approaches to solve the clogging problems (Bucks et al., 1979). The first approach is to focus attention on improving the quality of water before it reaches the emitters. The second approach is to develop emitters or devices to remain free from clogging. Extensive researches were done by Bucks et al. (1979) using 8 types of emitters and 6 different treatment processes for water from various sources. The study indicated that a combination of suitable physical and chemical treatments is needed to keep the emitters free from clogging. Particularly filtration along with pH adjustment was essential in order to control the chemical precipitation and bacterial growth. Furthermore, it showed that the expandable diaphragm emitters treated by chemicals for blockage problems, malfunction in its long term operations (Bucks et al., 1979).

In a hydraulic performance analysis on various kinds of emitters, Hezarjaribi et al., 2008 calculated the manufacturing variation coefficient, emitter discharge coefficient and emitter discharge exponent in order to establish flow sensitivity to pressure and compare manufacturers' specifications. The results indicated that for the chosen emitters the manufacturers supplied data are not reliable for design purposes. Reliable field tests are required prior to design of a drip system. In fact using the manufacturer's data will lead to non-uniformity of discharge throughout the system (Özekici and Sneed, 1995; Hezarjaribi et al., 2008).

Recent studies indicated that in developing countries the size of farms have been declining day-by-day under continuous population increase. Small land holders are not able to utilize modern technologies for those small farms. The efforts to find cheap and efficient method of irrigation for such small farms were investigated to raise the productivity without needing to use sophisticated technologies. In many of these works, microtube has been recommended as a low cost and easy to install emitters. Researchers indicated that the cost is substantially less than conventional emitter systems (Singh et al., 2009; Bhatnagar and Srivastava, 2003; Ella et al., 2009; Polak P, 1998).

When microtube is used as an emitter in a drip system this small tube itself dissipates energy to flow a certain discharge. The most important variable in its design is the calculation of energy losses due to friction at the inner wall of the tube and other minor components like entrance, exit, valves, bends, etc. These energy losses also represent the inlet pressure of the microtube since the outlet pressure is zero (Khatri et al., 1979; Bhuiyan, 1990)

Use of microtube as emitter was first conceptualised by Vermeiren and Jobling (1980). The sizes of the tubes were between 0.5 to 1.1mm and supposed to be the simplest and cheapest among all the water distributor devices. However, these small diameters are very susceptible to clogging.

Microtubes as small bore polyethylene tubes can be any sizes between 0.6mm and 4.0mm internal diameters. The discharge from a microtube varies according to the operating pressure, internal diameter and length of the tube. In other words, for a given internal diameter, the discharge of a microtube may be kept constant under various pressure conditions by adjusting its length. Therefore, if the pressure distribution along a lateral is known, uniform distribution of water can be achieved by using appropriate lengths of the microtubes. In this line of concept, Bhuiyan et al. (1990) proposed an algorithm to obtain a variable set of microtube lengths as an emitter to deliver uniform discharge along a typical lateral line. In their study a computational algorithm was developed for one lateral to simulate the set of microtube lengths and pressure head, microtube and lateral diameters and number discharge outlet.

Nowadays microtubes are widely used as an extension for micro-sprinkler or micro-jet systems to increase the outlet pressure and therefore to cover larger areas. These small tubes are also suitable for undulating lands where the pressure of the system varies considerably according to differences in elevation. Thus their lengths can be adjusted according to pressure heads to deliver a uniform discharge.

Recently International Development Enterprises (IDE), a non-profit organization based in Colorado, USA has suggested the use of this small diameter poly tubes in low cost drip irrigation kits. They believe that by connecting microtubes to plastic tape roll laterals and then to low-height tank would provide an easy to install and cheap low-pressure drip irrigation system, especially for developing countries in Africa and Southeast Asia (Ella et al., 2009). An experimental project was undertaken to use microtube as emitter to irrigate lands with various slopes. The results showed a wide variation in emission uniformity (EU) particularly with increasing uplands. The reason of this wide variation is due to the use of equal length microtubes throughout the system. In another similar study the elevated tank was provided to supply uniform discharge to different hilly terraces in northwest Himalayas (Bhatnagar and Srivastava, 2003). They recommended a star configuration layout of microtubes (1mm diameter) as emitters to service four rows by one lateral. Their results show that this configuration improved the emission uniformity to 94-98%. In effect it is found that the number of plants served by this configuration is very limited compared to commercial type of drip irrigation systems.

De Almeida et al. (2009) worked on an empirical approach to determine a new micro-sprinkler system where microtube lengths can be adjusted to deliver a nominal discharge for a given pressure. They found a set of pressure-discharge and pressure-length regression relationships for 1.07mm and 1.5mm microtubes to develop this new type of emitter. However, it is found that various flow regimes have been ignored for its head loss calculations.

Khatri et al. (1979) worked with seven different diameter microtubes (0.8 to 4 mm) to measure head losses in the drip irrigation system. The results showed that Darcy-Weisbach equation can be used along with Blasius friction factor equation for those hydraulically smooth microtubes. Analysis was done for the separation of minor losses and to produce coefficients for different flow conditions. However, finally it concluded that the microtubes are slightly smoother than the hydraulically smooth pipes as specified in the Moody diagram. Experimental results by Watters and Keller (1978) confirmed that the friction factor, f of the Darcy-Weisbach equation for smooth intermediate diameter pipes (4 to 12 mm) can be estimated by using Blasius formula. They provided some graphical solutions based on set of simple equations to represent a wide range flow rates encountered in drip irrigation systems. Experiments by von Bernuth and Wilson (1990) on larger diameter pipes (14, 16 and 26 mm) and for Reynolds number less than 100,000 also showed that the Blasius equation is an accurate predictor of the Darcy-Weisbach friction factors. It was concluded that Blasius equation has reasonable accuracy for estimating friction factor for a range of tubes in turbulent flow condition.

Schlichting (1968) used Nikurasde's (1933) data to illustrate the power relationship between the velocity distribution and Reynolds number and found the $1/7^{th}$ power law agress with the -0.25 exponent of the Blasius equation (1913). Hence, the Blasius equation is consistent with the power law for velocity distribution and has theoretical validity.

Experiments by von Bernuth and Wilson (1989) on larger diameter pipes (14, 16 and 26 mm) and for Reynolds number less than 100,000 also showed that a slightly modified Blasius equation can be an accurate predictor of Darcy-Weisbach friction factor.

In a similar investigation, Allen (1996) applied Churchill's equation (Churchill, 1977) for computing the Darcy-Weisbach friction factor for the full range of laminar, transition and fully turbulent flow regimes of the Moody diagram. The equation offers an explicit solution and performs better than the Blasius (1913), Colebrook and White (1937), Wood (1966), and Swamee and Jain (1976) equation.

However, the equation requires greater precision to compute its two different parameters with large exponents; one is primarily significant for large Reynolds numbers in the fully turbulent region and the other is significant at lower Reynolds numbers in the transition region.

Vekariya et al. (2011) carried out experiments using microtubes to determine discharges for various combinations of lengths (5-20 cm, diameters (1.2, 1.3, 1.5 and 2 mm) and operating pressures (1-11 m) to fit into a dimensionally homogenous equation. It is claimed that if the diameter of microtube is less than 1.2 mm, pressure dissipation occurred mostly due to the dominance of viscous forces.

Khatri et al. (1979) worked with seven different diameter microtubes (between 0.8 and 4 mm) to measure minor head losses in the drip irrigation system. Analysis was done for separation of minor losses and to produce coefficients for different flow condition. Additionally, the Blasius equation is simple and convenient to calculate the friction loss in a computer simulation. Unlike other complicated equations, Blasius can be formulated and rearranged to produce an equation with variables of flow rate, pipe length and diameter.

2.2. Design Considerations for Emitters along the Laterals

The concept of best emission uniformity was one of the key factors for the selection of lateral length and the number of emitters along it. In this regard, Christiansen (1942) carried out the first work for a pressurized sprinkler system. The paper proposed a coefficient of uniformity (U_c) for the system based on getting the measure on average of the absolute deviations of each observation from the mean observation. In equation format it can be written as:

$$U_c = 100\left(1 - \frac{\Delta q}{\bar{q}}\right) \tag{2.1}$$

where, $\Delta q = \frac{1}{n}\sum_{i=1}^{n}|q_i - \bar{q}|$, $\bar{q} = \frac{1}{n}\sum_{i=1}^{n} q_i$, and n = number of observed discharge values.

For a typical drip system, the emitter flow rates depend on emitter characteristics, aging and variability in manufacturing of emitters, friction head losses in the pipe network, topography of land and the number of clogged emitters in the system. Ideally, the application of water throughout the drip irrigation system should be absolutely uniform (Solomon and Keller, 1978). To achieve this objective Meyers and Bucks (1972) and Wu and Gitlin (1973) have developed a design procedure using different sized pipes. The theoretical performance of utilizing 5 different size emitters along a lateral line shows $\pm 3.3\%$ deviation from design discharge whilst for a single emitter size this value is +21% to -7.4% (Myers and Bucks, 1972). Howell and Hiller (1974) proposed a design procedure based on Christiansen coefficient of uniformity with considerations of emitter characteristics, frictional head losses and elevation differences of the topography. Wu and Gitlin (1977) proposed a technique to divide a lateral line into several sections and use different size pipes for each section. Their study shows that this simple fashion modifies the energy gradient curve of each section to a straight line except the last section. They also concluded when a lateral or manifold is divided into sections the mean discharge of each can be used to estimate the total energy drop by friction. This approach can be used for both uniform and non-uniform slopes; however, all slopes have to be down slopes.

Wu and Gitlin (1974, 1975) developed a dimensionless energy gradient line approach for direct calculation of energy drop for a range of emitter flow variation (less than 20%) along a single lateral line and a submain unit. Wu (1992) developed a computerised design technique using energy gradient line method to calculate the flow rates. A step-by-step (SBS) calculation has been carried out to compare the results and find out the deviations between two approaches. Karmeli and Peri (1977) suggested a design procedure for a single lateral by calculating the pressure head at each node backward from the downstream node.

Perold (1979) developed a graphical design procedure using computer program for the design of microirrigation pipe system. In this design, flow rate is assumed to be constant using adjustable microtube as feeder. The results of the simulation trials should be plotted to find an optimum size for each lateral. In another study, Perold (1977) developed a pocket calculator iterative procedure by means of a series of dimensionless solutions to design a constant size lateral with multiple outlets. The complement of the Christiansen coefficient is used as a criterion. Thus the average of the absolute deviations from the mean outflow $|\bar{\delta}|$ is considered as a design criterion. Finally a design chart is presented to find the permissible number of outlets for the given mean deviation $|\bar{\delta}|$.

Hathoot et al. (1993) developed a computer program that calculates the velocity head change and the variation of Reynolds number along the lateral pipe. Simple non-dimensional charts are prepared for some numerical examples to indicate that variation of Reynolds number along the lateral is important. In this study the emitter spacing is uniform and the outflow of individual emitters is evaluated stepwise from the first emitter. The uniformity of the system is the main design criteria to find the best lateral length.

Kang and Nishiyama (1996a, b) applied the finite element method and golden section search to find the operating pressure head of the lateral that can produce the required average emitter discharge. A computer program developed to calculate the best submain position and the operating pressure by input the known parameters such as average emitter discharge, required water emission uniformity and other conditions.

Through a different work, Vallesquino (2008) presented an alternative method to simulate hydraulic performance of laterals for sprinkler and trickle irrigation systems. The method estimates the outflow of similar type of laterals but with different lengths. In this study two head losses have been taken into account, one is transversal originating through the risers as a consequence of friction and local losses and another one is longitudinal originating along the lateral because of the friction and local losses. The results of model have been compared to step-by-step calculation (SBS) and energy gradient line version (EGL) and it is concluded that the results are closer to SBS method, while the deviation from EGL method is greater due to considering transversal head losses.

2.3. Pressure Distribution along the Laterals

According to Christiansen work (Christiansen, 1942) for sprinkler systems, a curve of pressure versus position along the lateral line seems to have the same general shape regardless the amount of flow rate, head loss and length of line. The flow condition in lateral line is steady and spatially varied with emitter outflows, thus the distribution of discharge along the line is increasing upstream. Due to the variation of discharge in the line, the energy gradient line will not be a straight line but a curve of the exponential type. Wu and Gitlin (1975) developed dimensionless pressure curves for three types of flow regimes in laterals. These dimensionless curves prompt the designer to calculate the energy drop along the lateral line. They formulated a general equation to calculate the total energy drop based on total discharge or average discharge in the lateral line. So the solution is only an approximation since the energy gradient line is determined by assuming all emitter flows are constant (or with small variation). To overcome the problem and increase the accuracy of design this direct Energy Gradient Line (EGL) approach modified into Revised Energy Gradient Line (REGL) approach using revised total discharge calculation. It is concluded that a mean flow approximation can be used to determine the adjusted total discharge for calculating total friction drop at the end of lateral line and used for direct calculation for emitter flows for the REGL approach. It is also concluded that the comparison of EGL and REGL with Step By Step (SBS) method for each emitter from downstream to upstream shows only 1% difference in emitter flow variation as long as the design is made within 20% emitter flow variation.

Keller and Karmeli (1974) confirmed that general shape and characteristics of the emitter head-loss curve are essentially independent of the emitter exponent and amount of head losses. A detailed analysis of many laterals using a computer model confirmed that for wide range of emitter characteristic values and pressure losses, the average pressure of the lateral occurs at 40% of the lateral line. They also found that approximately 77% of the total lateral head loss occurs between this 40% length of lateral line from starting point, leaving 23% between 60% of lateral line up to end.

Howell and Hiller (1974) presented design equations for the determination of lateral length to meet specific uniformity criteria. The emitter flow function is utilized to determine the allowable pressure loss to meet the uniformity standards. Emitter flow variation is a function of the uniformity coefficient. Thus, by knowing the emitter flow function, elevation change and design uniformity, the allowable pipe friction loss can be computed. Then taking the pipe size, pipe roughness coefficient, reduction coefficient for dividing flow, average emitter flow rate, allowable pipe friction loss determined previously, and either the number of emitters per lateral, the lateral length can be determined. For this purpose equations and dimensionless graphs developed to assist the design of laterals. Design input data includes the emitter flow function which could be obtained from manufacturers specification and reduction coefficient for dividing flow and emitter friction and inlet pressure of lateral. When the aforesaid data are known the dimensionless graphs can be used to determine lateral length as a function of the number of emitters per lateral.

Yildirim (2007) presented an analytical technique to solve hydraulic design problems of various types of multiple outlet pipe lines in different flow regimes and uniform line slope cases. This method could be generalized and applied for trickle, sprinkler and even gated pipes. To calculate the pressure and discharge the improved energy gradient line is determined based on the average friction drop with a simple exponential function to express the non-uniform outflow concept. To determine friction head losses, the Darcy-Weisbach formula is used and the kinetic head change is considered whereas minor head losses are neglected. To improve the accuracy of the technique, determination of pipe segments in different flow regimes have been taken into account. For the uniformity purposes, it is proposed that the maximum difference in the outlet operating pressure head along the pipeline is less than a percentage of the average pressure head. The result of some examples solved by this method are compared to numerical results of SBS method carried out by Hathoot et al. (1993), which shows fairly small difference.

2.4. Manifold Design

The design of the manifold is similar to the lateral design. However, the spacing between outlets is greater and larger flow rates are involved. The number of laterals and spacing between them determines the manifold length. The selection of the number of laterals depends on the considerations like: a) keeping within desired pressure differences, b) economic trade-off between the diameter and of the laterals and the manifold, c) the method of irrigation management, and d) the degree of automation of the system.

Most of laterals and manifolds in drip irrigation have been designed based on a single pipe size. The energy gradient line has been derived and presented by an exponential curve that is used as basis for designing laterals and manifold lines (Myers and Bucks, 1972; Wu and Gitlin, 1973). However, under certain field conditions, the length of laterals and manifolds may be relatively long and have non-uniform slopes, so the use of different diameter pipes is inevitable. Wu and Gitlin (1977a) proposed a graphical method to design lateral or manifold line with varying pipe sizes. They showed that by choosing different pipe diameters the energy gradient line will be close to the slope of the pipeline and therefore, the pressure variation will be reduced (Wu and Gitlin, 1977a).

3. Theoretical Development

3.1. Layout of a Typical Irrigation Subunit

Microtubes can be applied as emitters for undulating and hilly lands. Its length can be adjusted according to the pressure distribution along the lateral line and so equal discharge could be delivered to plants. Since the flow condition in the lateral line is steady and spatially varied having decreasing discharge in the downstream direction, the resultant energy grade line would follow an exponential curve.

Figure 3.1 shows a typical subunit where microtubes are emerging from the laterals to function as emitters. It can be seen that the microtubes have emerged from one side of the laterals so as to facilitate the operations of the farm machineries and to reduce the complexity of pipe layout. As can be seen the minimum pressure head required, at the end of the very last lateral (end-lateral) depends on the frictional and other minor head losses to overcome. The frictional head loss produced is directly proportional to the length of the microtube, which in turn is the distance between the lateral and the plant where water to be discharged. The choice of this distance would depend on the irrigator's practice and the physical facilities to be operated in the field.

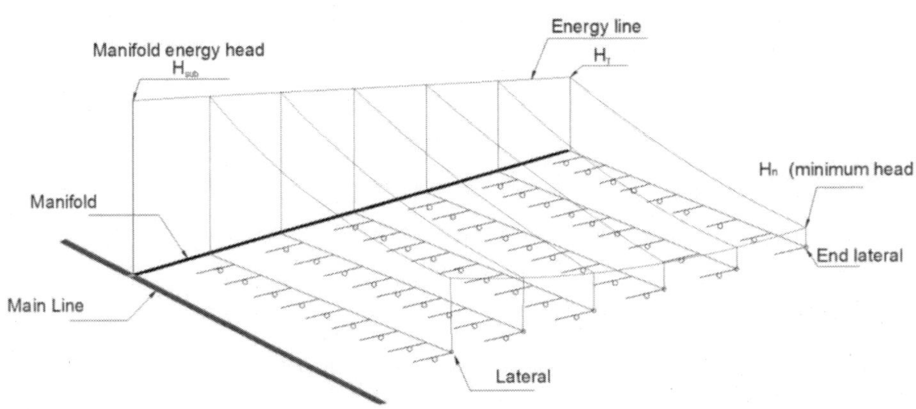

Figure 3.1. Schematic layout of a drip irrigation subunit using microtubes as emitters.

Figure 3.2 shows the details of a single lateral line where microtubes have varying lengths to compensate the extra heads developed along the lateral. It will be demonstrated later that to achieve equal discharges $q_1 = q_2 = ... = q_n$ for up (positive) and flat (zero) slopes, the microtube lengths would be $\ell_1 > \ell_2 > ... > (\ell_n = \ell_{\min})$. The microtube length ℓ_n at the end of the lateral is taken as the minimum length, determined from realistic distances between plants and laterals. The increase in microtube length is needed to dissipate difference of extra head (e.g., $H_{n-1} - H_n$) and give equal discharge for all the microtubes in the lateral. As shown in Figure 2 the increased length of microtubes can be wrapped around a stick (detail A), or the lateral itself (detail B), to keep a constant distance between lateral and the plant. In this study the term 'coil' is used with a diameter of 3 cm (using detail A), which can be altered by the designer in other circumstances. It is to note that there would have extra loss of energy due to these coiling.

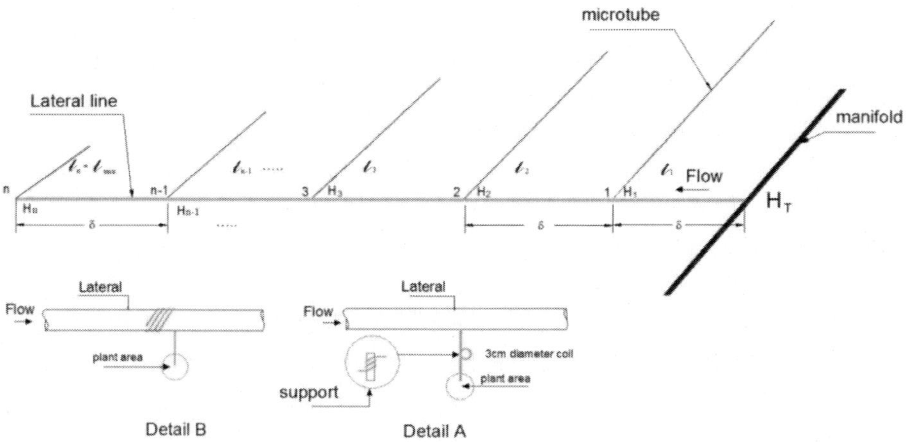

Figure 3.2. Schematic layout of a single lateral along with varying lengths of microtubes; details A and B shows the installation methods for making coils.

So with the given configuration, the varying length microtubes would release equal discharges throughout the end-lateral. To obtain these lengths and the number of coils, the total head at each microtube inlet should be calculated.

Subsequently there will be two layout configurations, one will have the same set of microtube lengths in the following laterals of the subunit, which in turn will require to limit the number of laterals in the subunit to produce a pre-defined emission uniformity (EU), and the other one is to vary the microtube lengths throughout the subunit to deliver equal discharges (EU = 100%). Let us name these two configurations, one will be called Pre-defined EU layout and the other one will be called Full EU layout. The conceptual setting of emission uniformity (EU) of these two layout configurations is discussed in more details in the following sections.

3.2. Basic Hydraulics

In this study Darcy-Weisbach equation is used to calculate the frictional losses in different pipelines throughout the subunit. As discharges along the lines are spatially varying, flow regimes are going to change according to the velocity conditions in it. Reynolds number may be calculated to know the flow regime and thereby to select the appropriate equations for estimating friction factors of Darcy-Weisbach equation. Darcy-Weisbach equation to calculate frictional head losses in pipes can be written in MKS units as

$$h_f = f \frac{l}{d} \frac{v^2}{2g} \quad (3.1)$$

For laminar flow the friction factor f can be written as

$$f = \frac{64}{R_e} \tag{3.2}$$

For turbulent flow with Reynolds number between 3000 and 100,000, Blasius equation gives good approximation for computing friction factor f , which can be written as

$$f = \frac{0.32}{R_e^{\ 0.25}} \tag{3.3}$$

where, R_e = Reynolds number, h_f = frictional head loss, ℓ and d = length and diameter of the pipes, g = acceleration due to gravity, and v = velocity of flow. Equations (3.1-3.3) can be combined to obtain the equations for laminar (Eq 3.4) and turbulent (Eq 3.5) flows, respectively:

$$h_f = \frac{1.32\ell q}{d^4} \tag{3.4}$$

$$h_f = \frac{0.486\ell q^{1.75}}{d^{4.75}} \tag{3.5}$$

where, h_f = frictional head loss (m), q = discharge (litre/hr), d = diameter of the pipe (mm), ℓ = length of the pipe (m). Kinematic viscosity of water at 15°C is taken as $\nu = 1.14 \times 10^{-6}$ m^2/s.

Velocity and other minor losses of the system can be written in general form as

$$h = k\frac{v^2}{2g} \tag{3.6}$$

where, k = head loss coefficient, which in three different minor loss coefficients are differentiated as: i) k_e = 1.2, to calculate entrance head loss assuming the entrance from lateral as a re-entrant one, ii) k_v = 1, to calculate velocity head, and iii) $k_c = c \times 1.3$, to calculate coil head loss, where 1.3 has been extrapolated (for $D/d \approx$ 12.0 and θ = 360°, where D and d are the coil and microtube diameters and θ is the angle of bend subtended at the centre) from Ito's diagram (Ito, 1960) on loss coefficient for smooth bends, c is the number of coils that can be computed from difference of two microtube lengths as $c = (l_{n-1} - l_n)/\pi D$. Only whole number of coils is taken for the calculation of head losses. Thus, Eq (3.6) can be rearranged to accommodate for the above three different minor losses as follows:

$$h_e = 0.0077\frac{q^2}{d^4} \tag{3.7}$$

$$h_v = 0.0064\frac{q^2}{d^4} \tag{3.8}$$

$$h_c = 0.0083m\frac{q^2}{d^4} \tag{3.9}$$

Energy grade line as shown in Figure 3.3 is related to head losses in one side of the lateral. Total head at the inlet of the microtube at point n can be calculated by summing all the head losses as follows:

$$h_e + h_v + h_f(n) = H_n \tag{3.10}$$

By placing the microtube with minimum length at the point n, the balance of energy heads between two successive points, $(n-1)$ and n can be written as

$$h_e + h_v + h_f(n-1) + h_c(n-1) = h_e + h_v + h_f(n) + h_{fl}(n) + S\delta \tag{3.11}$$

where, S and δ are slope of lateral and distance between microtubes, respectively. Since the discharges are same in all the microtubes, entrance and velocity head losses are equal in all the microtubes, so Eq (3.11) can be written as

$$h_f(n-1) + h_c(n-1) = h_f(n) + h_{fl}(n) + S\delta \tag{3.12}$$

By substituting full expressions for each of the head balance terms there will be a total four equations for four combinations of laminar and turbulent conditions in lateral and microtubes as follows:

1. Flow regimes are laminar in both the microtube and lateral

$$\frac{1.32\ell_{n-1}q}{d_m^{\ 4}} + \frac{0.0083cq^2}{d_m^{\ 4}} = \frac{1.32\ell_n q}{d_m^{\ 4}} + \frac{1.32nq\delta}{d_l^{\ 4}} + S\delta \tag{3.13}$$

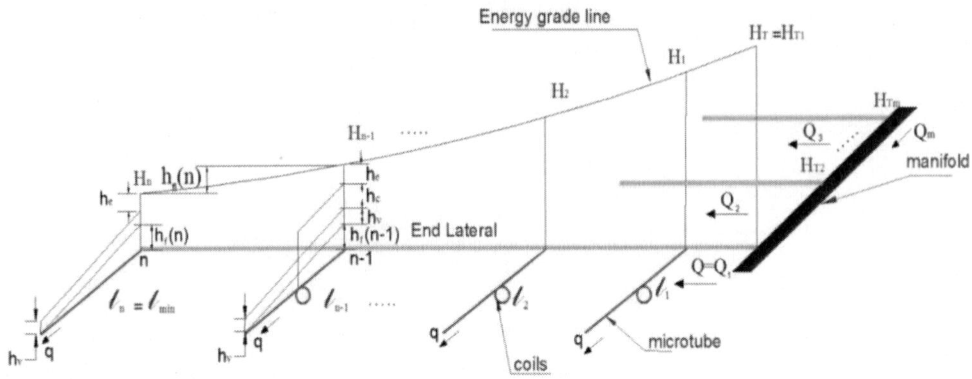

Figure 3.3. Energy grade line and head losses in one side of the lateral (h_e = entrance loss, h_v = velocity loss, h_c = coil head loss, $h_f(n)$ = microtube friction head loss at n, and $h_{fl}(n)$ = lateral friction head loss between and $n-1$).

2. Flow regimes are laminar and turbulent in microtube and lateral, respectively

$$\frac{1.32\ell_{n-1}q}{d_m^{\;4}}+\frac{0.0083cq^2}{d_m^{\;4}}=\frac{1.32\ell_n q}{d_m^{\;4}}+\frac{0.486nq^{1.75}\delta}{d_l^{\;4.75}}+S\delta \tag{3.14}$$

3. Flow regimes are turbulent and laminar in microtube and lateral, respectively

$$\frac{0.486\ell_{n-1}q^{1.75}}{d_m^{\;4.75}}+\frac{0.0083cq^2}{d_m^{\;4}}=\frac{0.486\ell_n q^{1.75}}{d_m^{\;4.75}}+\frac{1.32nq\delta}{d_l^{\;4}}+S\delta \tag{3.15}$$

4. Flow regimes are turbulent in both the microtube and lateral

$$\frac{0.486\ell_{n-1}q^{1.75}}{d_m^{\;4.75}}+\frac{0.0083cq^2}{d_m^{\;4}}=\frac{0.486\ell_n q^{1.75}}{d_m^{\;4.75}}+\frac{0.486nq^{1.75}\delta}{d_l^{\;4.75}}+S\delta \tag{3.16}$$

Therefore, the only unknown ℓ_{n-1} can be calculated directly from the above equations when the discharge required in the trees, diameters of microtube and lateral, slope, distance between microtubes, number of microtubes and the minimum length of microtube ($\ell_{\min}=\ell_n$) are all known parameters. Proceeding in this way up to the inlet of the end-lateral, all the microtube lengths will be known to deliver equal discharges q. After summing all the head losses along the lateral, the total head at the entry of the lateral is equal to inlet head H_T. So at the entry of the lateral, the total discharge would be $Q = nq$.

So it is apparent that the discharges through the microtubes are all equal in the end-lateral of a subunit. In the case of Pre-defined EU, the discharges through the subsequent laterals will vary ($Q_1 < Q_2 < \dots\dots < Q_m$, where m is the number of laterals obtained in the subunit) according to the total heads at their inlets. However, discharges through the microtubes in the following individual laterals will be equal (respective Qs will be divided by the number of microtubes) because of the same varying set of microtube lengths as of the end-lateral is in use. On the other hand, in the case of Full EU, discharges will be equal throughout the subunit because microtube lengths are varied according to the total heads developed in the respective inlets.

3.3. Pressure-Discharge Relationship

According to Keller and Karmeli (1974) and Howell and Hiler (1974) the power law relationship between flow rate and pressure head for emitters can be written as:

$$q = kH^x \tag{3.17}$$

where, q = discharge through emitters (litre/hr), k = the discharge coefficient that characterizes microtube emitters, H = pressure head at the entry of the emitters (m), and x = the discharge exponent that characterizes the emitter flow regime. For fully turbulent flow x = 0.5, for partially turbulent flow 0.5 < x < 0.7, for the unstable flow regime 0.7 < x < 1.0 and for laminar flow x = 1.0. For fixed orifice and nozzle type emitters, flow is always fully turbulent, so x = 0.5; for long path emitters $0.5 < x \leq 1.0$; and for pressure compensating emitters, $0.0 \leq x < 0.5$ (Keller and Karmeli, 1974).

In reality, discharges through emitters vary due to many reasons like pressure variation, manufacturers' imperfection, creeping over time, etc, as observed by many researchers (notably Bralts and Wu 1979b; Solomon 1979; Hezarjaribi et al. 2008). As an ideal design it is essential that the emitter flow variation be known, particularly since drip irrigation system efficiency depends on application uniformity and a successful system depends on physical and hydraulic characteristics of the emitters (Al-Amoud, 1995).

Manufacturing a set of emitters with the same k value is impossible due to their complexity and constraints (Solomon and Keller, 1978). Some of these constraints are related to mould damage and non-uniform mixing of raw materials during production process. Elastomeric materials are used to achieve flushing action and pressure compensation in the manufacture of pressure compensating emitters. These plastic parts are difficult to manufacture with consistent dimensions. Also, the resilient material may creep over a period of time and gradually change the flow rate even though pressure is constant (Solomon, 1979). Carpa and Scicolone (1998) indicated that the major sources of emitter flow rate variation are emitter design and the material used to manufacture the drip tubing and it's precision. Normally, manufacturers provide a coefficient k for each type of their products, which reflect the emitter's hydraulic characteristics. However, due to manufacturing variation and inconsistency, any two emitters of the same type tested at the same temperature and pressure can have different flow rates. As the flow rate of emitters is small, therefore any small variation within this device will cause large discharge variation through the system. As such, the manufacturing variation of different types of emitters has been tested to identify the manufacturers claimed coefficients (Solomon, 1977 and 1979; Solomon and Keller, 1978; Hezarjaribi et al. 2008).

By measuring the flow rates of a sample of emitters at a reference pressure head and then dividing the standard deviation of discharges by mean discharge, the manufacturer's coefficient of variation (CV) can be obtained. Typical values of this parameter may range from 0.02 to 0.1 for non-compensating emitters and even may go beyond 0.1 for some pressure compensating emitters.

In this study by applying microtubes as emitters in laterals, the problems of poor hydraulic design and non-uniform discharges due to manufacturers claimed coefficients, can be overcome. The coefficient k in this work is related to design factors like spacing between microtubes, diameters of lateral and microtubes, slopes, etc. Hence using the values according to the given configuration will give an accurate pressure-discharge relationship.

By using microtubes as emitters, the discharges of all the emitters in the lateral can be made equal (EU=100%) or varying with Pre-defined EU (say, > 90%) by applying Eqs (3.13-3.16), which in turn compose each lateral as a larger emitter attached to the manifold. On this basis, Eq (3.17) can be re-written for the characteristics of each lateral to relate discharge to pressure head of each lateral inlet as,

$$q = KH_T^{\ x} \tag{3.18}$$

where, q = flow rate through each microtube (litre/hr), K = the discharge coefficient that characterizes laterals, H_T = the pressure head at the lateral inlet (m) and x = the discharge exponent that characterizes the lateral flow regime.

In drip irrigation, ideally the application of water throughout the system should be uniform. It is necessary that the flow rates through the system should be uniform even though the pressure is not uniform (Solomon and Keller, 1978). In a well-designed drip irrigation system, the emission uniformity (EU) for emitters should be above a specific threshold level. The EU is a function of the expected discharge variation due to pressure variation throughout the system. Basically, EU is the ratio of the minimum emitter discharge to the average discharge of all the emitters under consideration, which can also be expressed as a percentage (Keller and Karmeli, 1974). An acceptable value of EU can be obtained by limiting the variation of pressure in the system. Limiting the pressure variation can decrease the variation of discharge in the emitters. Keller and Bliesner (1990b) recommended that EU should be at least 85% for drippers on flat terrain. Therefore, each lateral can be imagined as a larger emitter. This idea was first suggested by Keller and Karmeli (1974) and applied by Solomon and Keller (1978) for pressure distribution along the manifold.

In this research, two systems of design have been envisaged. In one system, called Pre-defined EU, the set of the microtube lengths calculated from end-lateral are provided for all the subsequent laterals unchanged to deliver varying discharges. The discharges of this system are varying in such a way so that the EU remains above certain specified threshold level. On the other system, called Full EU, microtube lengths are varied in such a way so that the discharges throughout the subunit become uniform to make EU equal to 100%.

3.4. Pre-defined EU System

In the Pre-defined EU system, the flow along the manifold line delivers a discharge to each lateral according to its inlet total head. Discharges from the manifold to the laterals follow the pressure-discharge relationship and energy grade line as shown in Figure 3.3. Since all parameters such as microtube lengths, slope and pipe diameters are constant in this design the discharges through the laterals vary according to pressure head along the manifold line and can be estimated using Eq (3.18). Therefore, the EU for the whole subunit can be computed according to Keller and Karmeli (1974) as

$$EU = 100\frac{Q_{ll}}{Q_{la}} \tag{3.19}$$

where, Q_{ll} = average of lowest ¼ of the lateral flow rates and Q_{la} = average of all the lateral flow rates in the system. To calculate EU of the system all the discharges of laterals (larger emitters) should be known. In the case of using traditional emitters, the main problems relating to estimation of discharges are due to pressure variation along the pipeline and manufacturer's non-uniformity in the batches of emitters produced.

Several studies have been carried out to estimate the discharges of emitters with respect to pressure variation. Some of these studies applied step-by-step method (Wu, 1992; Hathoot et al., 1993) and some others applied direct method to calculate discharges of the emitters (Vallesquino, 2008; Wu, 1992; Wu and Gitlin, 1973; Yildirim, 2007). Usually, the calculation of discharge is associated with some approximations to simplify the procedure and obtain a solution for the system. The most common approximation assumes that all outlets discharge nearly equal flow rates and based on this the frictional loss and energy gradient line could be obtained. The total frictional drop at a point and along the line can be calculated by an appropriate equation. Eventually, this method (Wu, 1992) offers a simple direct calculation of emitter flows based on energy gradient line (EGL). However, two types of errors are introduced by this simple EGL approach; one is associated with the shape of EGL and the other one is the total frictional drop at the end of the line.

The other approximation is the refinement of the previous method, by applying mean discharge of the emitters to develop a new formula to calculate total discharge of the lateral. Since the error produced by the EGL approach is caused mainly by the total discharge used to calculate the total frictional drop at the end of the line; rather we have to use actual discharge which is the summation of all the emitter discharges in the lateral line. Then this discharge could be applied for calculating total frictional drop of that line. The actual total discharge can be calculated by the mean flow of the emitters multiplied by the total number of emitters in the lateral line. The mean flow of the emitters can be determined from the mean pressure equation derived by Anyoji and Wu (1987) along the lateral line. Hence developing a formula combined with operating pressure and flow rate of the lateral is essential. However, this formula cannot be solved directly; a trial and error method is used to determine the total discharge in a subunit. The obtained total discharge is the revised total discharge and the energy gradient line that has been developed based on this value is called Revised Energy Gradient Line method (Wu, 1992).

In step-by-step technique (SBS) if the end pressure is given, the pressure and flow distribution for a given lateral can be found without resorting to approximations. Starting from the end, the outflow from an outlet, as determined by the pressure, is calculated. This gives the flow in the pipe section between this outlet and the next one, from which the pressure loss in this section can be found and thus the pressure at the next outlet (Perold, 1977). However, some field applications indicated that the manufacturer's information are not reliable and significant differences could occur particularly in high pressure systems (Hezarjaribi et al., 2008).

In this study as will be discussed subsequently, the stepwise method is applied to the laterals (as large emitters) for discharge calculation. The physical and mechanical characteristics of such large emitters are unvarying, and thus constant hydraulic performance is expected. It must be noted that when a set of microtube lengths $\ell_1, \ell_2, ..., \ell_n$ for certain design factors are chosen as emitters it can resolve mechanical issues related to manufacturing non-homogeneity of emitters. To find the hydraulic parameters x and K in the pressure-discharge relationship, regression method has been used. The performance of the large emitter (lateral) with several realistic discharges was tested and finally the best correlation between discharge and corresponding pressure heads obtained to use for design of the subunit.

3.5. Full EU System

To design a Full EU system with the highest emission uniformity (EU = 100%), it is required to vary the length of the microtubes for each laterals. The microtube lengths in successive laterals should be designed according to inlet pressure heads of the laterals. Figure 4 shows the schematic layout of a typical subunit where all the lateral flows are equal $Q = Q_1 = Q_2 = ... = Q_{m-1} = Q_m$. Considering the frictional losses of the manifold line, the lateral inlet heads can be as $H_{T1} < H_{T2} < ..... < H_{T(m-1)} < H_{Tm}$ using Eq (3.4) or Eq (3.5) according to flow regime.

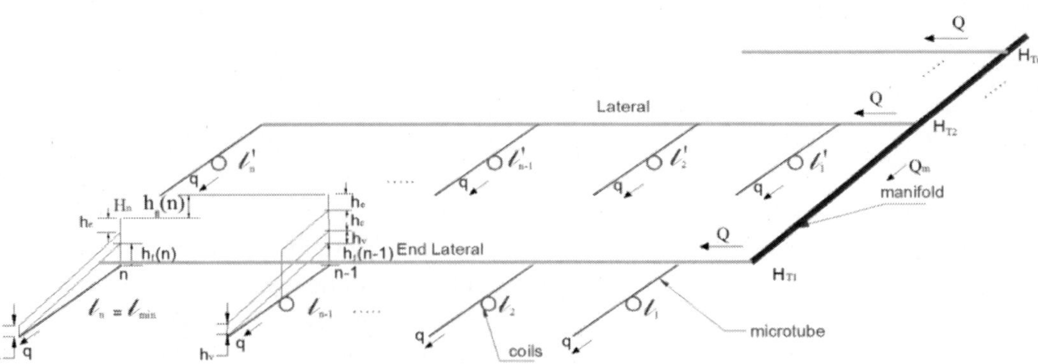

Figure 3.4. Schematic layout of a typical subunit with Full EU.

4. Model Development

4.1. Microtube Emitter Design

Since the calculation of pressures and microtube lengths are complicated and tedious, computer programs have been developed to simulate the system for different EU situations and given set of design parameters. The results of the step by step calculations, of the equations as explained in Section 3, are obtained by computer programs developed in Visual Basic 8.0. The required data for the design of the subunit are: lateral and microtube diameters d_l and d_m, microtube discharge requirement q, microtube spacing δ, and the ground slope along the pipeline S.

4.2. Design for Pre-defined EU

The design processes for Pre-defined EU system consists of three stages:

a) Lateral design - the program requires a requisite discharge to compute the lengths of the microtubes emerging from the laterals. As discussed before and according to head balancing equations in (3.13) to (3.16), the discharges for all the microtubes in the end-lateral are equal. The unknown variable of those equations is the microtube length at each outlet to dissipate extra frictional heads. So the total head at the inlet of the lateral (H_T) can be obtained by step-by-step calculations at different points of the microtubes up to the inlet of the lateral. The set of microtube lengths calculated at the end-lateral will be fitted in the subsequent laterals. As the total heads in the subsequent laterals are different, this set of microtube lengths is going to produce varying discharges through each lateral. However this unique set of microtube lengths obtained for a given suite of design parameters, would deliver equal discharges through these microtubes of the lateral considered.
b) Pressure-discharge relationship - an appropriate discharge range would be applied to the program to find the corresponding total heads. The choice of this discharge range depends on the practice by the irrigators and the pressure head that can be handled by the system. So a range of total heads would be estimated for the range of discharges through the microtubes of the end-lateral. The computed lateral heads and corresponding discharges are plotted and regression equations are obtained for application by the subsequent laterals. Since all the microtubes under a lateral deliver equal discharges, the lateral line discharge would be simple multiplication of this discharge with the number of microtube outlets in the lateral considered. Therefore, the pressure-discharge relationship obtained from the end-lateral could be expressed for all the laterals in the subunit.
c) Manifold design - to design the manifold line the program has been developed to calculate the frictional and other losses based on stepwise calculation from the end-lateral up to the inlet of the manifold. Then discharge of each lateral is calculated according to the pressure-discharge relationship obtained in end-lateral. At each step,

the Reynolds number along the manifold line is calculated and proper frictional head loss formulas are used to calculate the inlet heads of the laterals. The pressure head of the manifold and the number of laterals to achieve the best emission uniformity (say, EU > 90%) are the main outcome of the program.

4.3. Design for Full EU

It is desired to have a procedure with EU = 100% for the entire subunit. Unlike the above method, the Full EU system is based on variable microtube length for each lateral to dissipate extra frictional heads developed in the respective upstream reaches of the manifold. This Full EU system would follow the same procedure as explained in Pre-defined EU system except the set of microtube lengths calculated at the end-lateral will be used to calculate the new set of microtube lengths for the subsequent laterals.

4.4. Development of Model Algorithm

Three separate programs have been developed for aforementioned purposes as follow:

I. Program 1: Computing the microtube length and inlet total head of the end-lateral
II. Program 2: Find the best number of laterals to achieve the Pre-defined EU
III. Program 3: Finding the lengths of the microtubes for other laterals to achieve 100% EU in the entire system.

Program 1

In this program, the set of microtube lengths of the end-lateral are required to be calculated along with its total inlet head at the entry of the lateral. As such, the following parameters requirement to be decided prior to commencing simulation; these parameters can be varied according to design circumstances or field conditions:

1. Microtube diameter
2. Microtube minimum length, ℓ_n
3. Coil diameter
4. Lateral diameter
5. Lateral length
6. Microtube spacing, etc.

Through this program, the Eqs (3.13-3.16) have been applied for different flow conditions in the end-lateral. The flow regime from laminar to turbulent along the lateral line depends on the design discharge, spacing of microtubes and diameters. At this point, the conditional commands have been used in the program to compute the unknown microtube lengths. Figure 4.1 (flowchart) illustrates the computational algorithm for this code. The results of this program are arranged in graphs and tables to provide a range of discharges versus pressure heads for different diameters of microtube and lateral and microtube spacing. The regression curve to find the best relationship between pressure head and discharge has been undertaken and will be discussed later.

Program 2

This program takes the results from the previous program and then computes the frictional head losses along the manifold line. Using the same set of microtube lengths as in end-lateral and apply them in other laterals of the manifold, each lateral along the manifold can be imagined as large emitters. The hydraulic characteristics for these large emitters as explained before (Eq 3.18) are dependent on two parameters K and x which are obtained from Program 1. The microtube lengths and inlet pressure head of the end-lateral, manifold diameter, emission uniformity (EU) that the designers intend to apply and the number of laterals for the initiation of program to achieve a pre-defined EU are the major input data for this program. If the simulated EU from this program is less than the pre-defined EU, number of laterals was reduced for the next trial of the program. Figure 4.2 (flowchart) illustrates the computational algorithm for this code.

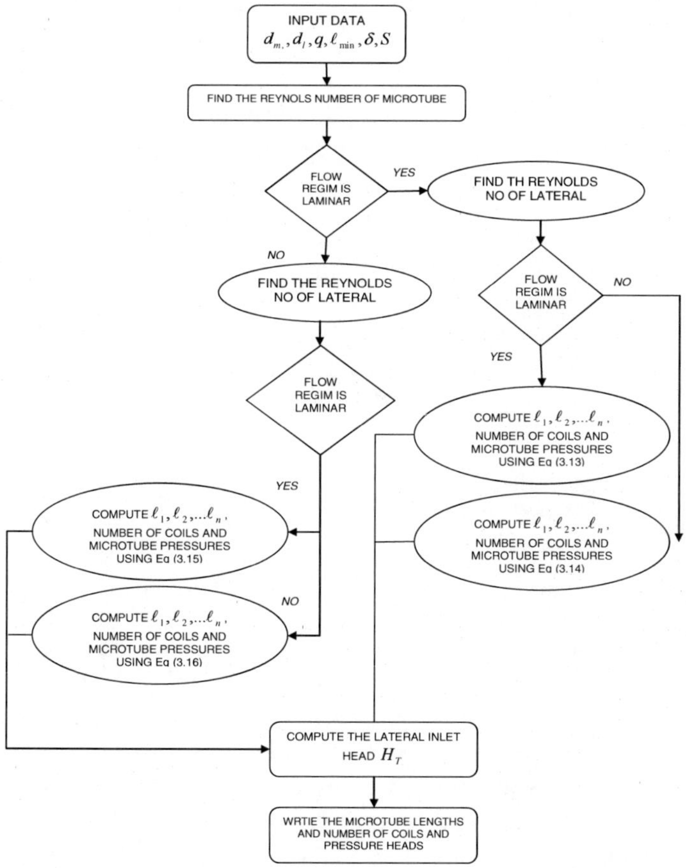

Figure 4.1. Flow chart of Program 1 to calculate the microtube lengths, inlet pressure head and number of coils in the end-lateral.

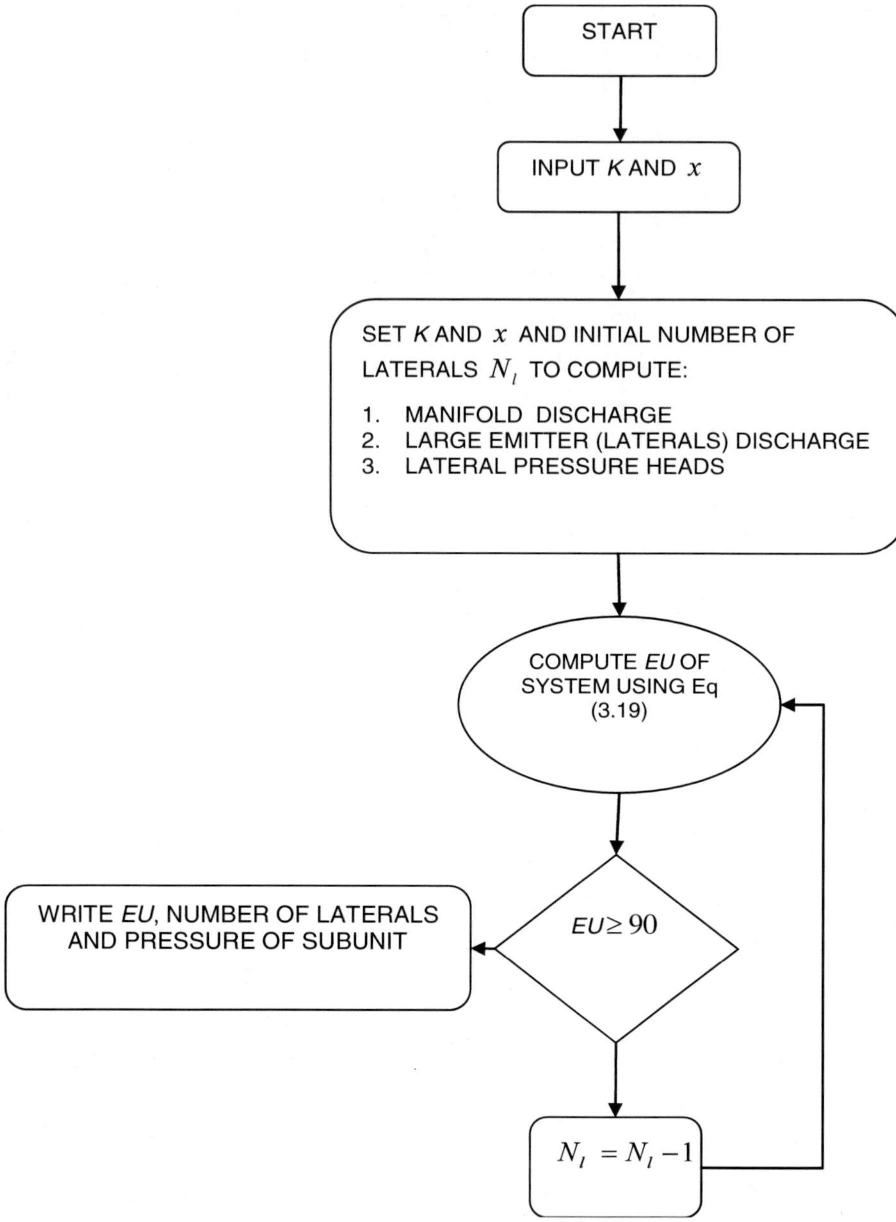

Figure 4.2. Flow chart of Program 2 to adjust the number of laterals required for a pre-defined EU.

This program starts with the total head of the end-lateral and adds the frictional head loss from the immediate upstream reach of the manifold to find the inlet pressure head of the following lateral. In each step Reynolds number was checked inside the manifold reach to select the correct frictional loss equation. So the new discharge in the following lateral is computed using Eq (3.18) and the process continued up to the first lateral of the manifold. The total number of laterals that would be sufficient for the given subunit can be obtained by adding or removing one or more new laterals each time and check whether it has reached the desired pre-defined EU. The output of this code includes the inlet pressure head of the manifold, total discharge and the best number of laterals applicable for the given subunit.

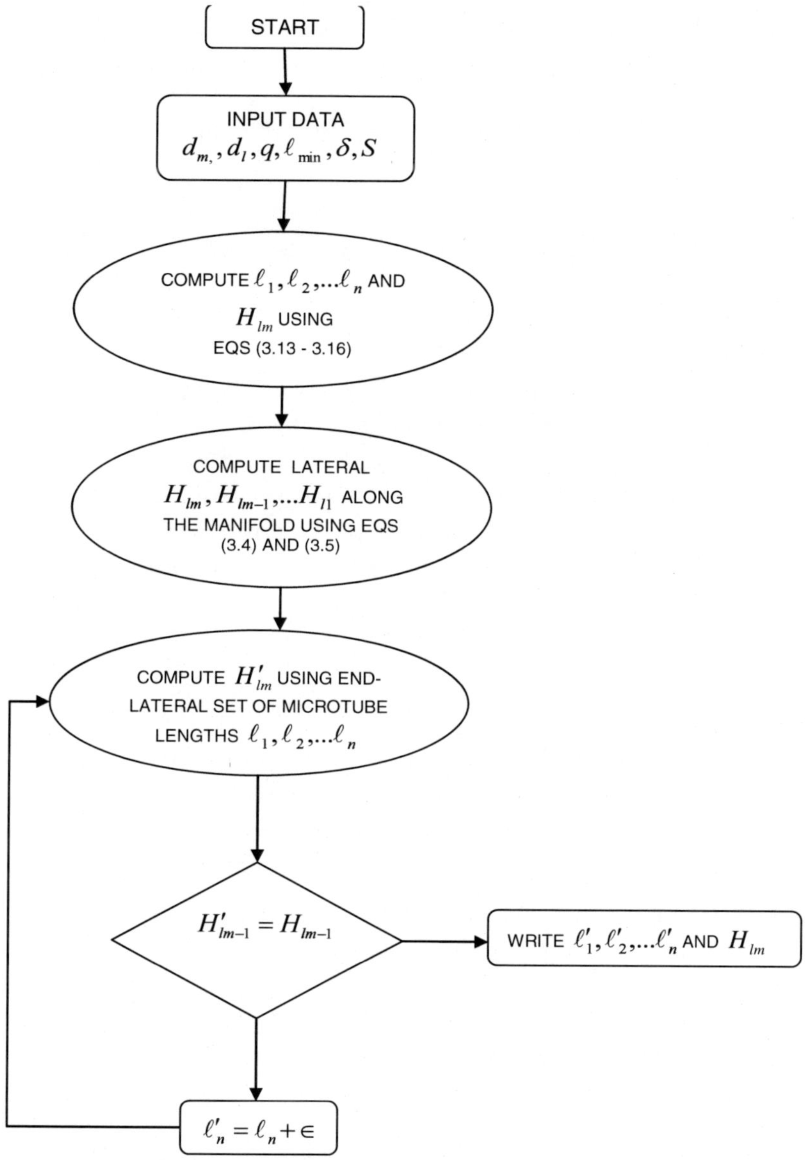

Figure 4.3. Flow chart of Program 3 to compute the new set of microtube lengths for successive laterals to achieve full emission uniformity.

Program 3

The main purpose of this code is to develop and design a system with 100% emission uniformity. The microtube lengths $\ell_1, \ell_2, ... \ell_{n-1}, \ell_n$ and the inlet total head of the end-lateral from Program 1 are applied for initiation of this program and then used to calculate the lengths of the microtubes for the following laterals. In this program each laterals are also imagined as large emitters with variable microtube lengths. The new set of lengths as the major part of this large emitter can be computed by increasing the lengths in trials of small increments to reach the value of the inlet pressure head of the lateral.

These new sets of microtube lengths are unique for each lateral. Subsequently for the convenience of design of the system, related graphs and tables are developed for direct calculation of the microtube lengths which has been discussed in the following chapter. Figure 4.3 shows the computational algorithm of this program.

5. RESULTS AND DISCUSSION

5.1. Microtube Design for the End-lateral

Since Program 1 has the main role in the design of the system and the results are to be used for the two other programs, some typical specifications of the system are chosen from a practical point of view for presenting the outcome. While the range of selected lateral diameters are taken as 10, 12, 14 and 16 mm, the microtube diameters are taken as 2, 3 and 4 mm from a practical consideration of market availability and also to keep them free from clogging. These microtubes are installed on one side of the laterals to discharge a given q to the roots of the plants. For flat terrain ($S = 0\%$) Table 5.1 shows the results related to inlet total head required (H_T), number of coils made (c) and the longest length (ℓ_{max}), and ℓ_i at $i = 5$ amongst all the estimated set of lengths of microtubes in that end-lateral. As can be seen in Table 5.1 the microtube lengths ℓ_{max} remain almost same as ℓ_{min} in d_m = 2 mm for the higher discharges, while total heads required has moved to high level.

To show the trend of microtube lengths in different terrains the program has also been run for two more slopes, S = 0.25% and 0.50%. The results show that some pressure heads for smaller diameter microtubes are very high and may be unrealistic for ordinary agricultural farms. These results have been shown in Appendix A.

5.2. Microtube Design for a Pre-defined EU

As discussed before in Program 2 the microtube lengths computed for end-lateral (in Program 1) would be assumed for all the following laterals of the subunit. Thus all laterals would have the same hydraulic characteristics and could be imagined to work as large emitters along the line of manifold. Evidently, flow rates of these emitters are not equal and can be estimated if the inlet total heads of each laterals are known along with the hydraulic constants (x and K) of Eq (3.18) are known. These hydraulic constants have been obtained for all the nominated combinations of microtube and lateral diameters. Figure 5.1(a-f) shows the power-law regression curves generated according to Program 1 results for pressures and discharges obtained for end-lateral. Different discharge ranges (q) are selected in those graphs to obtain the R^2 values greater than 0.999. It clearly illustrates that while the laterals are performed as emitters in its higher ranges, the smaller sized microtubes deliver less discharges with relatively higher heads, on the other hand larger sized microtubes deliver more discharges with relatively lower heads.

More graphs for other lateral diameters and slopes are given in Appendix B. The hydraulic constants and obtained from these graphs are summarized in Table 5.2. These computations are based on discharging water from one side of the lateral.

Table 5.1. Longest length microtube $\ell_{\max}$ (m), and ℓ_i, $i = 5$ (m), number of coils c and inlet pressure head required H_T (m) for the given microtubes and discharges (d_l = 10 mm, $\ell_{\min}$ = 1.25 m, n = 10, δ = 1 m and S = 0 %; the shaded cells are considered as unrealistic range of total heads for current scenario)

q litre/hr	d_m = 2 mm			d_m = 3 mm			d_m = 4 mm		
	H_T, m	$\ell_{\max}$ & (ℓ_5), m	c	H_T, m	$\ell_{\max}$ & (ℓ_5), m	c	H_T, m	$\ell_{\max}$ & (ℓ_5), m	c
1	1.12	1.31 (1.27)	1 (0)	0.24	1.58 (1.36)	4 (1)	0.1	2.32 (1.60)	11 (3)
2	2.47	1.31 (1.27)	1 (0)	0.54	1.56 (1.35)	3 (1)	0.21	2.24 (1.58)	11 (3)
3	4.04	1.3 (1.26)	1 (0)	0.87	1.54 (1.34)	3 (0)	0.34	2.18 (1.56)	10 (3)
4	5.83	1.3 (1.26)	1 (0)	1.25	1.52 (1.34)	3 (0)	0.48	2.13 (1.54)	9 (3)
5	7.84	1.3 (1.26)	1 (0)	1.67	1.51 (1.33)	3 (0)	0.63	2.08 (1.52)	9 (2)
6	10.06	1.29 (1.26)	1 (0)	2.14	1.49 (1.33)	3 (0)	0.8	2.03 (1.51)	8 (2)
7	12.49	1.29 (1.26)	0 (0)	2.65	1.48 (1.32)	3 (0)	0.99	2 (1.50)	8 (2)
8	15.15	1.29 (1.26)	0 (0)	3.19	1.47 (1.32)	2 (0)	1.19	1.96 (1.48)	8 (2)
10	21.07	1.29 (1.26)	0 (0)	4.41	1.45 (1.31)	2 (0)	1.6	1.9 (1.46)	7 (2)
12	27.9	1.28 (1.26)	0 (0)	5.81	1.42 (1.31)	2 (0)	2.07	1.78 (1.45)	6 (2)
15	39.79	1.27 (1.26)	0 (0)	8.19	1.38 (1.30)	1 (0)	2.85	1.66 (1.42)	4 (1)
18	53.67	1.27 (1.26)	0 (0)	10.99	1.36 (1.30)	1 (0)	3.79	1.62 (1.41)	4 (1)
20	78.86	1.26 (1.25)	0 (0)	13.05	1.36 (1.29)	1 (0)	4.47	1.6 (1.38)	4 (1)

a) d_m =4mm, q =1-8 (litre/hr)

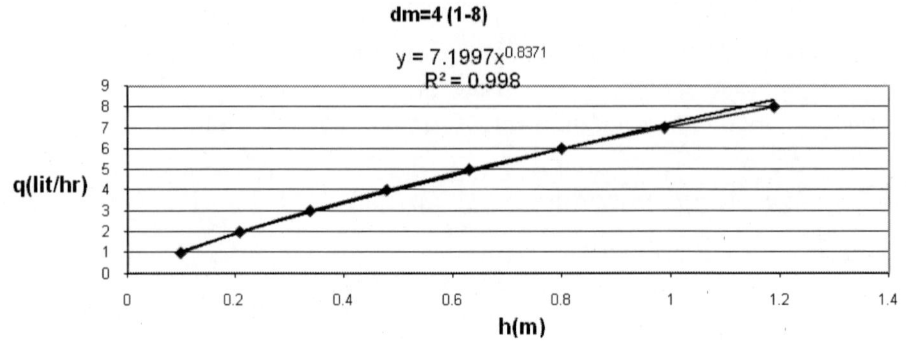

b) d_m =4mm, q =8-20 (litre/hr)

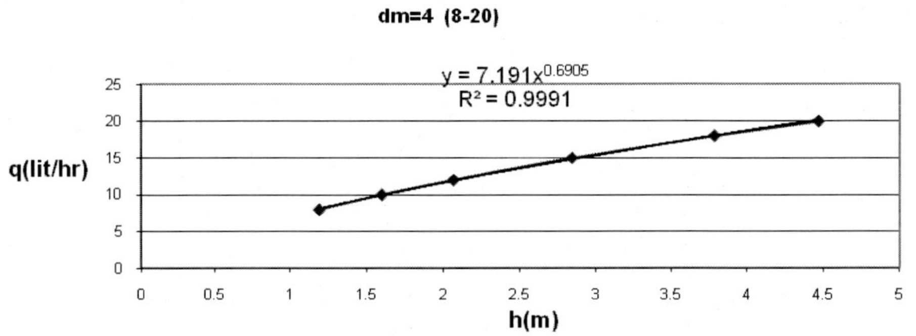

c) d_m =3mm, q =1-6 (litre/hr)

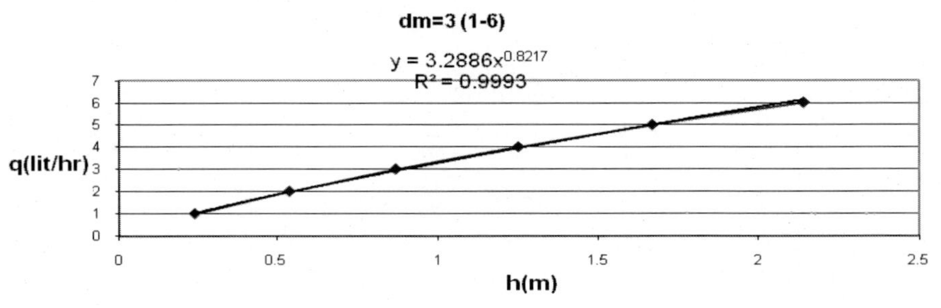

d) d_m =3mm, q =6-20 (litre/hr)

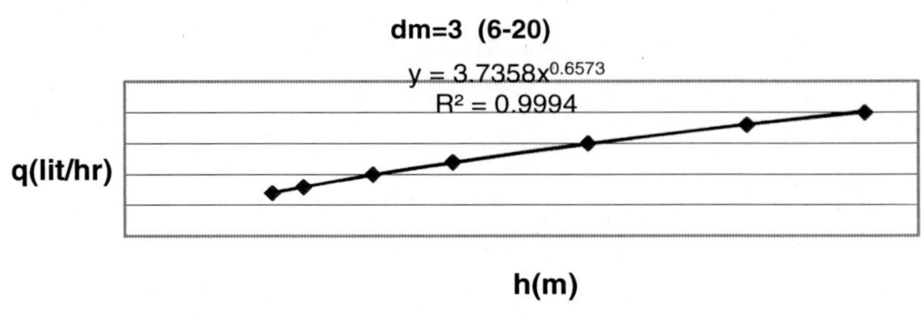

e) d_m =2mm, q =1-6 (litre/hr)

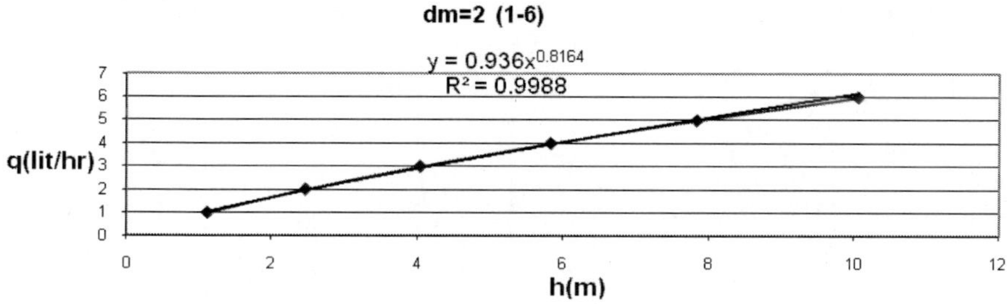

f) d_m =2mm, q =6-20 (litre/hr)

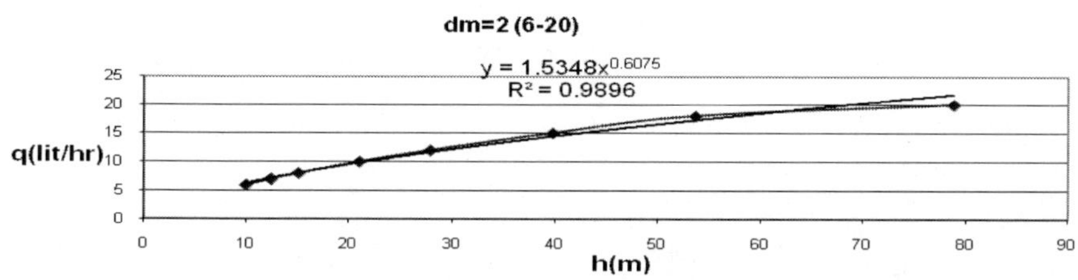

Figure 5.1. Pressure-discharge relationships for laterals as large emitters (d_l = 10 mm, ℓ_{min} = 1.25 m, n = 10, δ = 1 m and S = 0 %).

The flow rate from individual emitters depends on operating pressure, water temperature, manufacturing variations and degree to which the emitters are plugged. It is usually assumed that over some range of pressure (H), the emitter flow rate (Q) is proportional to H^x. The value of x characterizes the flow regime of the emitter, which normally ranges between zero and one (Solomon, 1985). In this study, the manufacturing variation and plugging, which were the main obstacles for improvement of emission uniformity have been eliminated due to the use of microtubes. As laterals can now be imagined as large emitters along the manifold, it is evident that the operating total heads would be the main effective variable for discharges along the outlets of the laterals. In the following tables the input data employed in some examples demonstrate the application of this approach. In a previous work by Hathoot et al. (1993) based on step by step method, the average emitter discharge, the corresponding average pressure head, the exponent and number of emitters are given so the individual discharge and pressure head for each emitter can be calculated. Similar technique has been applied for this study using laterals as large emitters with pressure-discharge relationship as shown in Table 5.2.

Table 5.2. Pressure-discharge relationships for different laterals with ℓ_{min} = 1.25m, n = 10, and δ = 1 m; a) for S = 0%, b) for S = 0.25%, and c) for S = 0.50%

a) $S = 0\%$

d_l	d_m	2 mm		3 mm		4 mm	
10 mm	q, litre/hr	6-20	1-6	6-20	1-6	8-20	1-8
	H_T (m)	7.8-78.8	1.1-7.84	2.1-3.2	0.24-2.1	1.2-4.4	0.1-1.2
	x	0.6075	0.8164	0.6573	0.8217	0.6905	0.8371
	K	1.53	0.94	3.73	3.28	7.19	7.20
12 mm	q, litre/hr	6-20	1-6	6-20	1-6	8-20	1-8
	H_T (m)	9.95-76.2	1.1-9.95	2.0-12.8	0.23-2.0	1.05-4.2	0.08-1.05
	x	0.6116	0.8171	0.6341	0.8203	0.6617	0.8093
	K	1.52	0.94	3.94	3.44	7.81	7.97
14 mm	q, litre/hr	6-20	1-6	6-20	1-6	8-20	1-8
	H_T (m)	9.92-73.7	2.2-9.9	2.0-12.6	0.22-2.0	0.99-4.1	0.07-0.99
	x	0.6048	0.8153	0.6438	0.8131	0.6436	0.7842
	K	1.57	0.95	3.94	3.50	8.10	8.28
16 mm	q, litre/hr	6-20	1-6	6-20	1-6	8-20	1-8
	H_T (m)	9.9-71.1	1.1-9.9	1.9-12.6	0.22-1.9	0.97-4.0	0.07-0.97
	x	0.6239	0.8155	0.6470	0.8160	0.6373	0.7926
	K	1.47	0.9517	3.93	3.54	8.24	8.50

b) $S = 0.25\%$

d_l	d_m	2 mm		3 mm		4 mm	
10 mm	q, litre/hr	6-20	1-6	6-20	1-6	8-20	1-8
	H_T (m)	10.9-78.9	1.28-10.9	2.9-14.8	0.41-2.9	2.3-6.25	0.26-2.3
	x	0.6304	0.8352	0.7814	0.902	0.9177	0.9506
	K	1.37	0.83	1.559	2.30	3.69	3.73
12 mm	q, litre/hr	6-20	1-6	6-20	1-6	8-20	1-8
	H_T (m)	10.8-76.3	1.27-10.8	2.8-14.8	0.39-2.8	2.1-6.25	0.24-2.1
	x	0.6216	0.8351	0.7272	0.8953	0.8546	0.9439

	K	1.43	0.8421	2.82	2.38	4.06	3.96
14 mm	q, litre/hr	6-20	1-6	6-20	1-6	8-20	1-8
	H_T (m)	10.8-73.7	1.26-10.8	1.5-6.4	0.38-2.8	2.7-6.4	0.23-2.1
	x	0.6399	0.8348	0.7195	0.8910	0.8211	0.9353
	K	1.34	0.84	2.86	2.42	4.31	4.07
16mm	q, litre/hr	6-20	1-6	6-20	1-6	8-20	1-8
	H_T (m)	10.7-71.2	1.2-10.7	2.8-15	0.38-2.8	2.0-6.5	0.23-2.0
	x	0.7160	0.8354	0.7190	0.8927	0.7985	0.9408
	K	2.89	0.84	2.89	2.43	4.45	4.12

c) S = 0.50%

d_l	d_m	2 mm		3 mm		4 mm	
10 mm	q, litre/hr	6-20	1-6	6-20	1-6	8-20	1-8
	H_T (m)	11.7-79.06	1.4-11.7	3.8-16.6	0.57-3.8	3.4-8.04	0.42-3.4
	x	0.6526	0.8556	0.819	0.9403	1.0701	0.9897
	K	1.23	0.74	1.99	1.73	2.09	2.42
12 mm	q, litre/hr	6-20	1-6	6-20	1-6	8-20	1-8
	H_T (m)	3.7-76.4	1.4-11.7	3.7-16.9	0.55-3.7	3.3-8.31	0.4-3.29
	x	0.6557	0.8531	0.7941	0.935	0.9579	0.9857
	K	1.23	0.7577	2.11	1.79	2.55	2.53
14 mm	q, litre/hr	6-20	1-6	6-20	1-6	8-20	1-8
	H_T (m)	11.6-73.8	1.4-11.6	3.7-17.2	0.54-3.7	3.2-8.7	0.4-3.2
	x	0.6612	0.8541	0.7739	0.9325	0.917	0.9897
	K	1.21	0.75	2.2	1.81	2.69	2.56
16 mm	q, litre/hr	6-20	1-6	6-20	1-6	8-20	1-8
	H_T (m)	11.6-71.3	1.4-11.6	3.6-17.5	0.5-3.6	3.2-9	0.39-7.5
	x	0.6661	0.8523	0.7622	0.9347	0.8789	0.9836
	K	1.19	0.7626	2.25	1.82	2.86	2.60

5.3. Microtube Design for Full EU

Energy Gradient Line Analysis

In Program 3, microtubes and laterals are designed to discharge equal flow rates throughout the subunit. When the inlet total head at the end-lateral is known, all the frictional head losses of the respective manifold reaches can be added together to give corresponding inlet total heads of the following laterals. Indeed the end-lateral can be imagined as a large emitter with known hydraulic characteristics as explained earlier. Since it is obvious that all the lateral flow rates are equal, the frictional losses along the manifold are based on the respective flow rates in the reach, which are simple multiple of the corresponding number of laterals downstream. For convenience of design the energy gradient lines for laterals and manifold can be developed to calculate the total heads and the resulting microtube lengths required directly.

The energy gradient line is a curve of exponential type for any lateral or manifold. Since flow rate decreases in the downstream direction, it is obvious that frictional losses in the upper reaches are much higher than the lower reaches. The total head losses are computed from total length of the pipe and its discharge, and other minor losses. In this study other minor losses are due to coils, entrance and exit velocity losses.

Wu and Gitlin (1975) and Wu (1992) worked to utilize the energy gradient line approach for direct calculation of the emitter discharges. The approach evolved a method for direct calculation of outflows based on an assumption that there would have 10-20% flow variation along the lateral. Evidently, the flow in the lateral is steady and spatially varied with decreasing discharges in the downstream direction. For a given pipe, the energy drop can be expressed as

$$\frac{dH}{dx} = -kQ_x^{\ m} \tag{5.1}$$

where, k and m are constants for a given flow condition (m = 1 for laminar flow, m =1.75 for turbulent flow in smooth pipes and m =2 for turbulent flow; m =1.852 if Hazen-Williams equation is applied), dH = the energy drop for a given length dx and Q_x = the discharge at a section of length x measured from the inlet point. Assuming the drip irrigation system is designed for an emitter with little or almost no variation in flow, the shape of the head drop ratio curve can be derived mathematically (Wu and Gitlin, 1975) from Eq (5.1) as:

$$R_i = 1 - (1 - i)^{1+m} \tag{5.2}$$

where, $R_i = \Delta H_i / \Delta H$ is a head drop ratio along a lateral, ΔH = total head drop up to the end of the lateral, ΔH_i is the total head drop at a length ratio i, $i = x / X$ is a length ratio for a given position x from the lateral inlet to the total lateral length X. Wu and Gitlin (1975) defined Eq (5.2) for three dimensionless curves based on m = 1, 1.75 and 2 for laminar, turbulent (in smooth pipe) and fully turbulent flows, respectively. In their approach, they considered frictional losses only, neglected minor losses from the emitters and laterals. Later Wu (1992) developed equations and computer program, where emitters with turbulent and non-turbulent flows were made to calculate the emitter flow rates directly. These approaches are developed based on the assumption that the variation of discharges through the emitters is negligible. Step-by-step method was in use to estimate the frictional losses of the laterals.

The simulated curves that have been drawn in Figure 5.2(a-c) for head drop ratios, $\Delta H_i / \Delta H$ versus length ratios, i (to be called simulated energy gradient lines) along the length of the laterals are showing some degrees of deviation from the energy gradient curves using the right-hand-side of Eq (5.2). The right-hand-side of the Eq (5.2) is said to be applicable to laminar, turbulent and fully turbulent flows if the exponent values are taken as m = 1, 1.75 and 2, respectively. Evidently the deviations as shown in Figure 5.2(a-c) are due to the existence of mixed flow regimes and minor losses considered in the simulations of the head drop ratios. The right-hand-side of the Eq (5.2) is developed only for frictional head loss for a given flow regime. As such the head drop ratio curves in Figure 5.2(a-c) would match to Eq (5.2) if the exponent values are changed to m = 0.32, 0.1 and 0.05 for microtube diameters d_m = 4, 3 and 2 mm, respectively.

To compare the energy gradient lines for different microtube sizes the simulation results have also been plotted in Figure 5.3(a-d) for two laterals 12mm and 16mm and slopes $S = 0\%$ and 0.25%. As shown in these figures, energy gradient lines for the flat terrain are very close and more or less straight, while in upslope terrain the variation of these dimensionless curves is wide.

It is perceptible that the selection of the sectional energy slope ($\Delta H_i / \Delta x_i$) should be such that the simulated energy gradient line is always above the head required along the pipe line. A steeper slope simulated energy gradient line (particularly at the upstream-end) will produce smaller sized pipes, which will have less cost (Wu, 1975). It is said that the optimal shape of the simulated energy gradient line is a curve having sag above the straight line around $0.15\Delta H$ at the middle of the pipeline profile ($i = 0.5$). However comparing the cost of the design determined by using a straight energy gradient line with the one using optimal energy gradient line, the cost difference is only about 2.5% (Wu, 1975). So from a given set of design parameters on discharges (q), diameters (d_m, d_l), spacing (δ) and total head drops (ΔH), a choice based on straight energy gradient line as shown in Figure 5.3(a-d) would be sufficient to decide a combination of these parameters (q, d_m d_l, δ, and ΔH) in different field situations (S %). Visual inspection of these figures shows that the design combination close to the straight energy gradient line would produce an optimum system. For flat terrain, it could be almost any combination since most curves are close to straight line. For sloped lands it shows that the microtube size 3mm with any discharge between q = 3 and 8 litre/hr would have better performances.

a) for d_m = 4mm

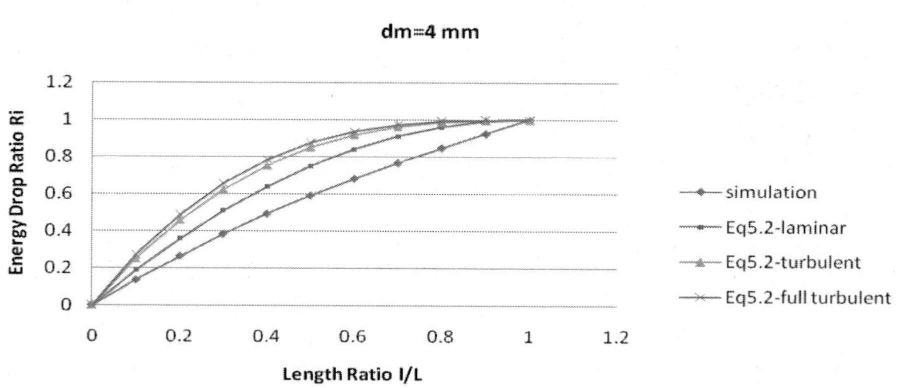

b) for d_m = 3mm

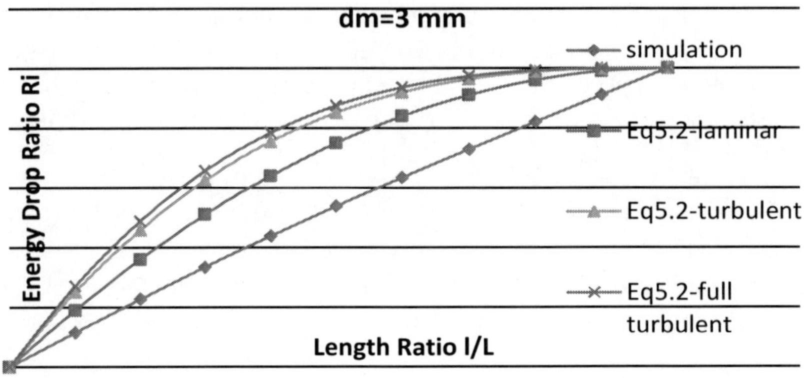

c) for d_m = 2mm

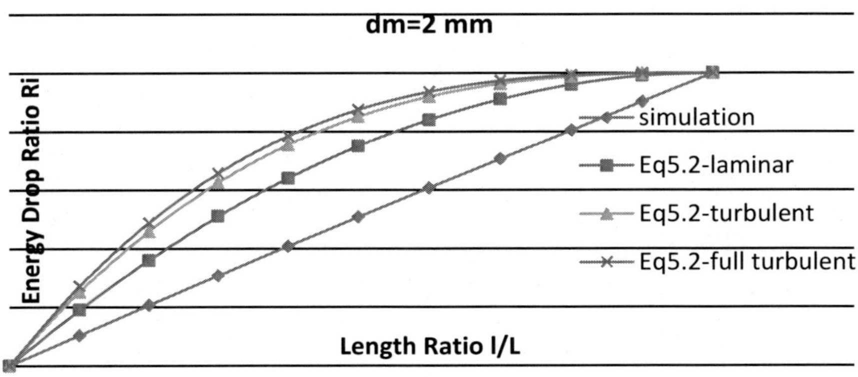

Figure 5.2. Comparison between simulated ($\Delta H_i / \Delta H$ vs i) and Eq (5.2) derived energy gradient lines.

a) d_l = 12 mm, S = 0%

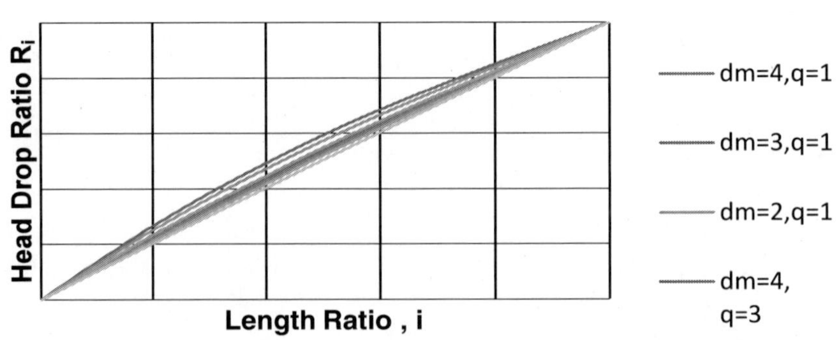

b) d_l = 16 mm, S = 0%

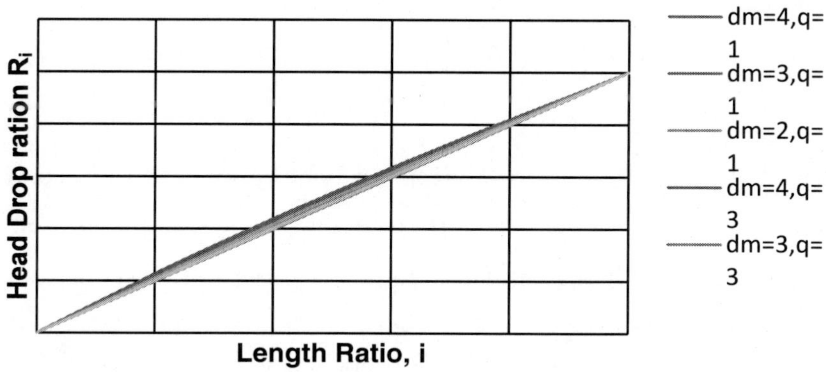

c) d_l = 12 mm, S = 0.25%

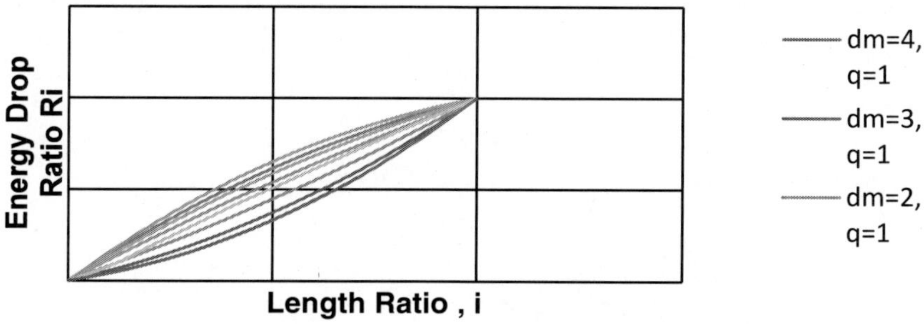

d) d_l = 16 mm, S = 0.25%

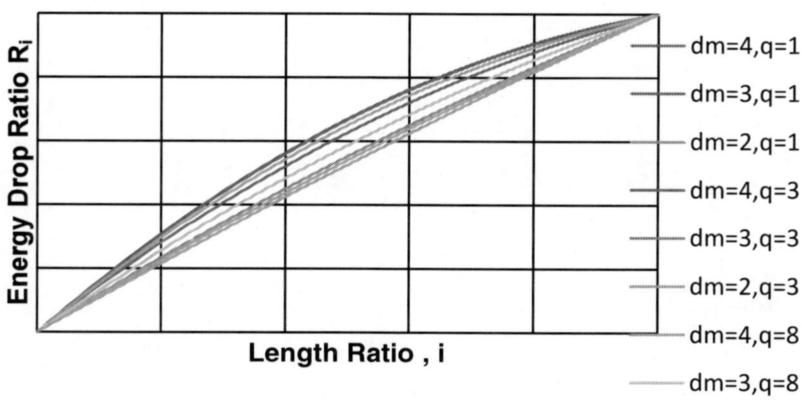

Figure 5.3. Simulated energy gradient lines for typical laterals; a) for d_l = 12mm, S = 0%, b) d_l = 16 mm, S = 0%, c) d_l = 12 mm, S = 0.25% and d) d_l = 16 mm, S = 0.25% (there are 10 trees placed at an interval of 1m space along the laterals).

Microtube Length Computation

A technique has been developed for calculating the microtube lengths without the use of the program. A solution associated with design tables and graphs has been developed for designers to compute the varied microtube lengths. Developing the energy gradient line and generalizing its concept to the calculation of microtube length differences, have prompted to define a new parameter, M_i as microtube length difference ratio

$$M_i = \Delta\ell_i / \Delta\ell \tag{5.3}$$

where, $i = x/X$ is a length ratio for a given position x from the lateral inlet to the total lateral length X, $\Delta\ell_i = (\ell_{max} - \ell_i)$ is microtube length difference at i and $\Delta\ell = (\ell_{max} - \ell_{min})$ is the total length difference for a given lateral. The simulation indicates that the microtube total length difference $\Delta\ell$ varies according to the flow conditions in the microtubes and laterals. The $\Delta\ell$ values shown in Figure 5.4(a-f) indicate that it reduces with smaller diameter microtubes along a given lateral. It is shown that the values of $\Delta\ell$ are declining with increasing discharge capacity of the microtubes and becoming almost constant for smaller diameter microtube (say at d_m = 2mm). The $\Delta\ell$ values also reduce with the increase of lateral sizes.

a) d_l = 12mm, S = 0%,

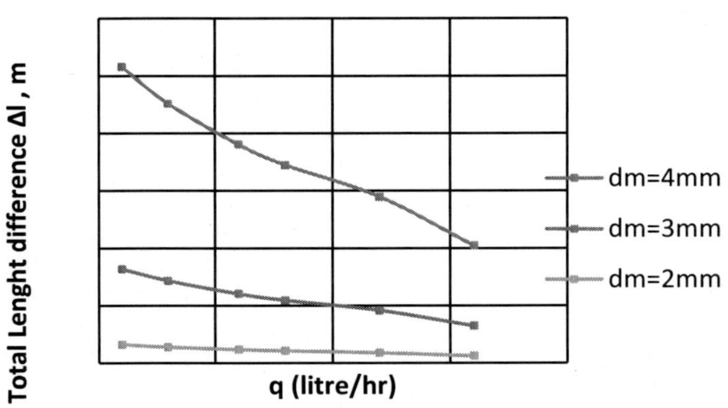

b) d_l = 14mm, S = 0%,

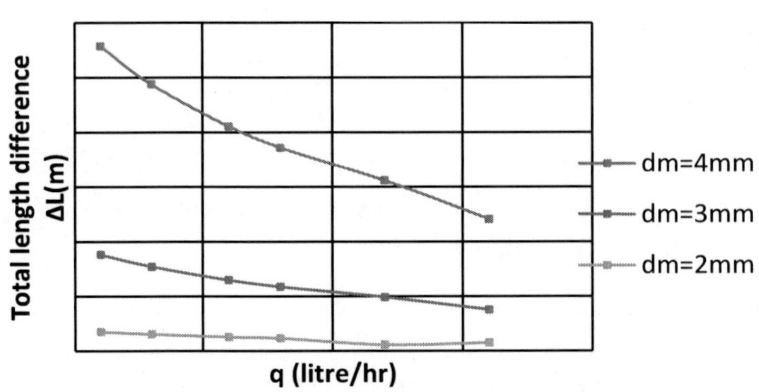

c) d_l = 16mm, S = 0%,

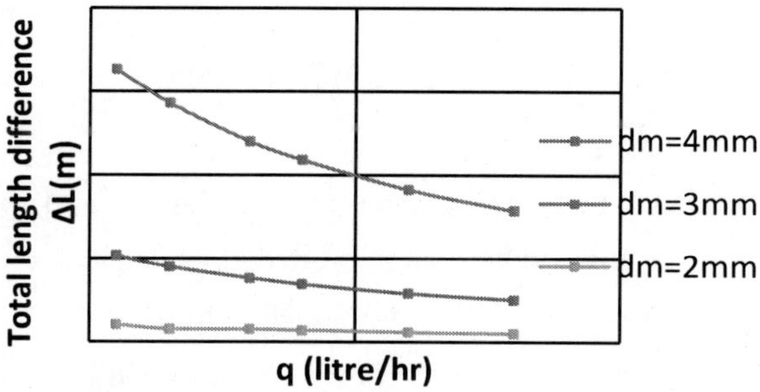

d) d_l = 12mm, S = 0.25%

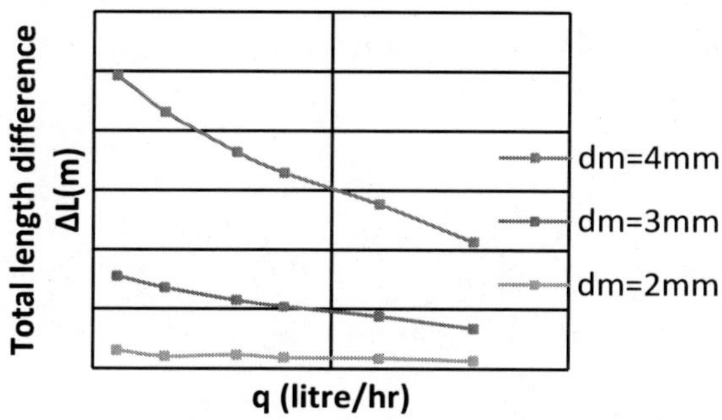

e) 14mm, 0.25%,

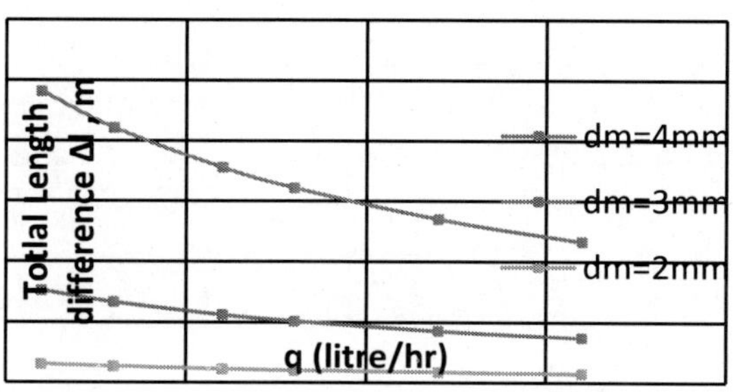

f) d_l = 16mm, S = 0.25%,

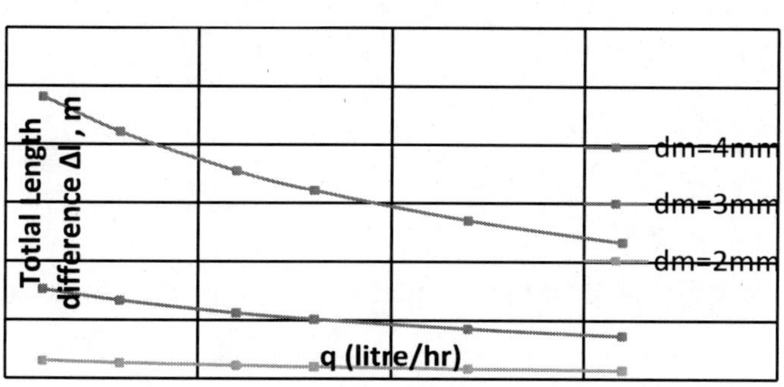

Figure 5.4. Microtube total length difference $\Delta\ell$ for typical laterals; a) for $d_l =$ 12mm, $S =$ 0%, b) for $d_l =$ 14mm, c) for $d_l =$ 16mm $S =$ 0%, d) for $d_l =$ 12mm, $S =$ 0.25%, e) for $d_l =$ 14mm, $S =$ 0.25% and f) for $d_l =$ 16mm, $S =$ 0.25% (there are 10 trees placed at an interval of 1m space along the lateral).

Subsequently it is found that with all the possible lateral and microtube diameters and discharges, the simulated microtube length difference ratios (M_i) follow a general exponential trend (shown in Figure 5.5). This general trend only varies with respect to the land slope, S. While M_i can be read in Figure 5.5 at a position i, we need to know $\ell_{\min}$ for a given lateral in order to know the whole set of microtube lengths along that lateral. Analysis undertaken shows that the minimum microtube lengths, $\ell_{\min} = \ell_n$ for the laterals can be expressed as a general quadratic equation as

$$\ell_{\min} = a \times j^2 + b \times j + c \tag{5.4}$$

where, $j = p/P$ is the position ratio for laterals along the manifold line at p from the manifold inlet to the total manifold length P, and a, b and c are constants as given in Table 5.3(a-b) for two lateral sizes $d_l =$ 12mm and 16mm, respectively. As microtube length-difference distribution is related to head drop distribution, it is possible that Eq (5.2) can also be written for length difference ratios M_i in a similar form but with a different exponential value m' (Eq 5.5) in order to calculate microtube lengths for any laterals. The simulated length differences curves (Figure 5.5) that are applicable for any discharges and diameters of lateral and microtube are based on all the friction and minor head loses considered earlier, fits well if $m' =$ 1.2 and 1.64 for slopes of $S =$ 0% and 0.25%, respectively.

$$M_i = 1 - (1 - i)^{m'+1} \tag{5.5}$$

where, M_i as above mentioned can be computed by knowing $\Delta\ell$ and $\ell_{\min}$ of each lateral. So by substituting Eq (5.5) into Eq (5.3) we can get microtube lengths at different i in a given lateral as

$$\ell_i = \ell_{\min} + \Delta\ell(1 - i)^{m'+1} \tag{5.6}$$

where ℓ_i is the microtube length at the position i, $\Delta\ell$ is the total length difference obtained from Figure 5.4(a-f), and $\ell_{\min}$ is the minimum microtube length, for a particular lateral position j in the manifold, obtained from Eq (5.4) and Table 5.3(a-b). These tables can be developed for a wide range of possible design parameters so that the design can be done without utilizing the computer programs directly. To calculate the operating pressure head of the subunit, hence the pressure head of the end lateral obtained from Figure 5.5 should be added with the frictional head losses along the manifold reaches to give H_{sub}.

The results of the simulation also indicate that the microtube total length differences for each of the laterals in the manifold is equal to constant; in fact, the microtube variation is a constant value despite of its varied length in successive laterals along the manifold.

Table 5.3(a-c) show the coefficient of for two lateral slopes simulated for 6 different flow rates.

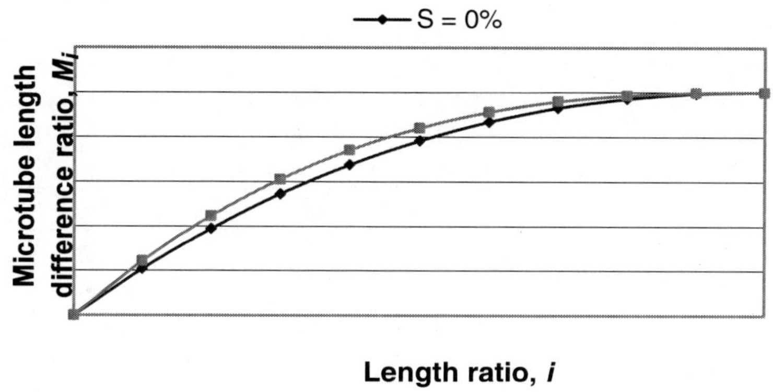

Figure 5.5. Microtube length difference distribution for any discharge, and microtube and lateral size (this graph is derived for 10 trees placed at an interval of 2m space along the lateral).

Table 5.3. Coefficients of Eq (5.4) for minimum possible lengths of microtube $\ell_{\min}$; a) for d_l =12mm and b) for d_l =16 mm (these tables are derived for laterals at an interval of 1m space along the manifold, where 10 trees are placed at an interval of 1m space along each of the lateral)

a) d_l =12mm

q (litre/hr)		d_m = 2mm			d_m = 3mm			d_m = 4mm		
		a	b	c	a	b	c	a	b	c
1	S=0%	0.0100	-0.0190	0.5125	0.0530	-0.1137	0.5788	0.1613	-0.3062	0.7010
	S=0.25%	0.0093	-0.0200	0.5126	0.0509	-0.1066	0.5659	0.1613	-0.3388	0.7072
3	S=0%	0.0012	-0.0104	0.5105	0.0225	-0.0650	0.5556	-0.0790	-0.0207	0.6368
	S=0.25%	0.0020	-0.0111	0.5100	0.0095	-0.0554	0.5505	0.0341	-0.1836	0.6635
6	S=0%	0.0096	-0.0202	0.5141	0.0479	-0.0995	0.5734	0.1589	-0.3033	0.7038
	S=0.25%	0.0100	-0.0210	0.5129	0.0327	-0.0937	0.5671	0.1600	-0.3389	0.7081
8	S=0%	0.0120	-0.0227	0.5142	0.0620	-0.1157	0.5715	0.2054	-0.4344	0.7838
	S=0.25%	0.0118	-0.0250	0.5150	0.0604	-0.1268	0.5754	0.2054	-0.3815	0.7378
12	S=0%	0.0157	-0.0303	0.5172	0.0970	-0.1804	0.5969	0.2891	-0.5498	0.8015
	S=0.25%	0.0157	-0.0303	0.5172	0.0970	-0.1804	0.5969	0.2891	-0.5498	0.8015
16	S=0%	0.0200	-0.0380	0.5195	0.0812	-0.1937	0.6107	0.3614	-0.6875	0.8466
	S=0.25%	0.0200	-0.0380	0.5195	0.0812	-0.1937	0.6107	0.3614	-0.6875	0.8580

b) d_l =14mm

q (litre/hr)		d_m = 2mm			d_m = 3mm			d_m = 4mm		
		a	b	c	a	b	c	a	b	c
1	S=0%	0.018	-0.0609	0.5469	0.018	-0.0609	0.5469	0.1612	-0.306	0.6753
	S=0.25%	0.01	-0.019	0.5109	0.0504	-0.096	0.5558	0.01	-0.019	0.5109
3	S=0%	0.0189	-0.0589	0.5451	0.0189	-0.0589	0.5451	0.0559	-0.1862	0.6446
	S=0.25%	0.0039	-0.0118	0.509	0.0039	-0.0118	0.509	0.0559	-0.1862	0.6446
6	S=0%	0.0552	-0.1043	0.5584	0.0522	-0.0983	0.5554	0.5191	-1.0233	1.0381
	S=0.25%	0.0093	-0.0181	0.5106	0.0325	-0.084	0.5574	0.1591	-0.3288	0.6985
8	S=0%	0.518	-1.0485	1.0356	0.518	-1.0485	1.0356	0.2254	-0.4032	0.7115
	S=0.25%	0.012	-0.0227	0.5344	0.0616	-0.1153	0.5633	0.205	-0.3868	0.711
12	S=0%	0.0511	-0.1395	0.6121	0.0511	-0.1395	0.5904	0.2738	-0.5335	0.786
	S=0.25%	0.0163	-0.0308	0.5163	0.0818	-0.1555	0.5826	0.3045	-0.5494	0.7867
16	S=0%	0.1423	-0.2369	0.6064	0.1423	-0.2369	0.6064	0.3616	-0.6877	0.8393
	S=0.25%	0.0204	-0.0385	0.5192	0.1423	-0.2369	0.6064	0.382	-0.7098	0.8479

c) d_l =16mm

q (litre/hr)		d_m = 2mm			d_m = 3mm			d_m = 4mm		
		a	b	c	a	b	c	a	b	c
1	S=0%	0.01	-0.019	0.5101	0.0496	-0.0963	0.5518	0.1613	-0.3068	0.6628
	S=0.25%	0.0112	-0.021	0.5109	0.0534	-0.0991	0.5518	0.0112	-0.021	0.5109
3	S=0%	0.0479	-0.0921	0.5497	0.0479	-0.0921	0.5497	0.0596	-0.1901	0.6337
	S=0.25%	0.0039	-0.0118	0.5082	0.0039	-0.0118	0.5082	0.0559	-0.1862	0.6322
6	S=0%	0.0091	-0.0185	0.5104	0.0479	-0.0921	0.5497	0.1591	-0.3033	0.6657
	S=0.25%	0.0091	-0.0179	0.5098	0.0557	-0.1046	0.5544	0.1591	-0.3033	0.6657
8	S=0%	0.0616	-0.1152	0.5593	0.0616	-0.1152	0.5593	0.2255	-0.4032	0.6995
	S=0.25%	0.0121	-0.0227	0.5117	0.0830	-0.1383	0.5644	0.0559	-0.1862	0.6322
12	S=0%	0.0159	-0.033	0.518	0.0818	-0.1555	0.5787	0.2893	-0.5501	0.7748
	S=0.25%	0.0159	-0.0306	0.5156	0.0818	-0.1555	0.5787	0.2893	-0.5501	0.7748
16	S=0%	0.0202	-0.0383	0.5188	0.1421	-0.2366	0.6048	0.3820	-0.7098	0.8433
	S=0.25%	0.0202	-0.0383	0.5188	0.1218	-0.2262	0.6079	0.3616	-0.6877	0.8347

6. CONCLUSION

To reduce the problem of clogging and blockage of emitters due to sediment and solvable materials in drip irrigation, microtubes with diameters 2-4 mm have been proposed to act as emitters. To carry out a well-organized design with the maximum emission uniformity (EU) within the system, two solution algorithms, one is on Pre-defined EU and another one on Full EU, have been developed employing microtubes as emitters along the laterals.

Microtubes with different lengths were designed to deliver equal volumes of water according to pressure head distribution along the lateral. The design starts from end-lateral with the calculation of a set of microtube lengths to flow a given discharge. A program has been developed to calculate these microtube lengths and corresponding total heads. It needs information about the microtube number and spacing, microtube and lateral diameters, and slope to flow a given discharge. For these design information, pressure-discharge power relationships are generated for subsequent application. The length of the microtube is a variable to adjust the pressure and deliver steady and uniform flow rates through the reaches of laterals, manifolds, subunits and main lines.

6.1. Design Using Pre-defined Emission Uniformity (EU)

In the Pre-defined EU, another simple computer code was developed to estimate the number of laterals that can be installed in the manifold to fulfil the EU requirement set by the designer. It needs information about the lateral spacing and diameter, manifold diameter and land slope to calculate total head developing in different reaches of the manifold. This code takes input on microtube lengths and pressure-discharge relationships from the first program of the end-lateral. The set of microtube lengths are replicated in all the following laterals. Hence, laterals work as large emitters to deliver resultant discharges for the varying pressure heads in the manifold reaches. Therefore, the number of laterals obtained from this code can be fitted in a manifold to perform in the best EU measure set by the designer. The Pre-defined EU analysis shows that using smaller diameter microtube increases the number of lateral that can be implemented in the manifold. However the constraint of applying the smallest microtube size (d_m = 2mm) is to increase total head requisite of the system and thus the operational cost. On the other hand, the larger microtube size (d_m = 4mm) needs longer microtube lengths when lower discharges are to be delivered. Considering all these prospects and constraints, the middle sized microtube (d_m = 3mm) could be a balance to fulfil the present requirement of irrigation.

6.2. Design Using Full Emission Uniformity (EU)

In the other design using Full EU algorithm, the microtube lengths are varied by some increment in each successive lateral compared to end-lateral microtubes. The variation of these microtube lengths are estimated according to the pressure heads to be dissipated in each lateral. Energy gradient line approach as is applied to lateral and manifold lines to show the variation of total head, is analogised with the variation of microtube lengths to present a new technique for design. Hence, by developing design tables, graphs and regression equations, the technique illustrates a simple design procedure for this approach. The challenge of compensating the variable pressure heads in the network has been made by obtaining the proportionate difference of microtube lengths required (simulated) in a balance to the extra total head there compared to the position at microtube ℓ_{min} of the end-lateral. The simulation result shows that the exponential energy gradient curve can be matched with the microtube length difference distribution curve applicable for wide ranges of discharge and microtube and lateral sizes. The Eq (5.6) along with a graph on total difference of microtube lengths Figure 5.4(a-f) and the minimum microtube length required Table 5.3(a-c) for a given lateral size, can make the design for uniform distribution of water in the subunit, much easier.

APPENDIX A

Longest length microtube $\ell_{\max}$ (m), number of coils and inlet pressure head required (m) for the given microtubes and discharges (d_l = 10, 12, 14, 16 mm, $\ell_{\min}$ = 1.25 m, = 10, δ = 1 m and S = 0, 0.25 and 0.5 %; the shaded cells are considered as unrealistic range of total heads and coil numbers)

Table A.1 $S = 0\%$

a) d_l = 12 mm

q litre/hr	d_m = 2 mm			d_m = 3 mm			d_m = 4 mm		
	H_T, m	$\ell_{\max}$ & (ℓ_5), m	c	H_T, m	$\ell_{\max}$ & (ℓ_5), m	c	H_T, m	$\ell_{\max}$ & (ℓ_5), m	c
1	1.11	1.28 (1.26)	0	0.23	1.41 (1.30)	2	0.08	1.76 (1.42)	5
2	2.44	1.28 (1.26)	0	0.5	1.4 (1.30)	2	0.18	1.73 (1.41)	5
3	3.99	1.27 (1.25)	0	0.82	1.39 (1.29)	2	0.29	1.7 (1.40)	5
4	5.76	1.27 (1.25)	0	1.18	1.38 (1.29)	1	0.41	1.67 (1.39)	5
5	7.75	1.27 (1.25)	0	1.59	1.37 (1.29)	1	0.55	1.65 (1.38)	4
6	9.95	1.27 (1.25)	0	2.04	1.37 (1.29)	1	0.7	1.63 (1.37)	4
7	12.38	1.27 (1.25)	0	2.53	1.36 (1.28)	1	0.87	1.61 (1.37)	4
8	15.03	1.27 (1.25)	0	3.06	1.35 (1.28)	1	1.05	1.59 (1.37)	4
10	20.98	1.26 (1.25)	0	4.26	1.34 (1.28)	1	1.45	1.56 (1.36)	3
12	27.8	1.26 (1.25)	0	5.63	1.34 (1.28)	1	1.89	1.53 (1.35)	3
15	39.6	1.26 (1.25)	0	8.01	1.31 (1.27)	1	2.66	1.45 (1.34)	2
18	53.54	1.26 (1.25)	0	10.77	1.3 (1.27	1	3.5	1.43 (1.33)	2
20	76.22	1.25 (1.25)	0	1.3	1.3 (1.27)	1	4.22	1.41 (1.32)	2

(b) d_l = 14 mm

q litre/hr	d_m = 2 mm			d_m = 3 mm			d_m = 4 mm		
	H_T, m	$\ell_{\max}$ & (ℓ_5), m	c	H_T, m	$\ell_{\max}$ & (ℓ_5), m	c	H_T, m	$\ell_{\max}$ & (ℓ_5), m	c
1	1.1	1.26 (1.25)	0	0.22	1.33 (1.27)	1	0.07	1.52 (1.34)	3
2	2.43	1.26 (1.25)	0	0.49	1.33 (1.27)	1	0.16	1.51 (1.33)	3
3	3.97	1.26 (1.25)	0	0.8	1.32 (1.27)	1	0.27	1.49 (1.33)	3
4	5.73	1.26 (1.25)	0	1.16	1.32 (1.27)	1	0.38	1.47 (1.32)	2

q litre/hr	d_m = 2 mm			d_m = 3 mm			d_m = 4 mm		
	H_T, m	ℓ_{max} & (ℓ_5), m	c	H_T, m	ℓ_{max} & (ℓ_5), m	c	H_T, m	ℓ_{max} & (ℓ_5), m	c
5	7.72	1.26 (1.25)	0	1.55	1.31 (1.27)	1	0.52	1.46 (1.32)	2
6	9.92	1.26 (1.25)	0	2	1.31 (1.27)	1	0.66	1.45 (1.31)	2
7	12.34	1.26 (1.25)	0	2.48	1.31 (1.27)	1	0.82	1.44 (1.31)	2
8	14.99	1.26 (1.25)	0	3.01	1.3 (1.26)	1	0.99	1.43 (1.31)	2
10	20.93	1.26 (1.25)	0	4.19	1.3 (1.26)	1	1.38	1.41 (1.30)	2
12	27.76	1.25 (1.25)	0	5.56	1.29 (1.26)	1	1.82	1.4 (1.30)	2
15	39.64	1.25 (1.25)	0	7.9	1.29 (1.26)	0	2.58	1.38 (1.29)	1
18	53.49	1.25 (1.25)	0	10.6	1.28 (1.26)	0	3.45	1.35 (1.29)	1
20	73.69	1.25 (1.25)	0	12.66	1.27 (1.26)	0	4.11	1.34 (1.28)	1

c) d_l = 16 mm

q litre/hr	d_m = 2 mm			d_m = 3 mm			d_m = 4 mm		
	H_T, m	ℓ_{max} & (ℓ_5), m	c	H_T, m	ℓ_{max} & (ℓ_5), m	c	H_T, m	ℓ_{max} & (ℓ_5), m	c
1	1.1	1.25 (1.25)	0	0.22	1.30 (1.26)	1	0.07	1.41 (1.30)	2
2	2.42	1.25 (1.25)	0	0.48	1.29 (1.26)	1	0.16	1.4 (1.30)	2
3	3.96	1.25 (1.25)	0	0.79	1.29 (1.26)	0	0.26	1.39 (1.29)	2
4	5.7	1.25 (1.25)	0	1.14	1.29 (1.26)	0	0.37	1.38 (1.29)	1
5	7.71	1.25 (1.25)	0	1.54	1.29 (1.26)	0	0.5	1.37 (1.29)	1
6	9.91	1.25 (1.25)	0	1.97	1.28 (1.26)	0	0.64	1.37 (1.29)	1
7	12.33	1.25 (1.25)	0	2.45	1.28 (1.26)	0	0.8	1.36 (1.28)	1
8	14.97	1.25 (1.25)	0	2.97	1.28 (1.26)	0	0.97	1.35 (1.28)	1
10	20.91	1.25 (1.25)	0	4.15	1.28 (1.26)	0	1.34	1.34 (1.28)	1
12	27.73	1.25 (1.25)	0	5.5	1.27 (1.25)	0	1.78	1.34 (1.28)	1

Table A.1. (Continued)

q litre/hr	d_m = 2 mm			d_m = 3 mm			d_m = 4 mm		
	H_T, m	ℓ_{max} & (ℓ_5), m	c	H_T, m	ℓ_{max} & (ℓ_5), m	c	H_T, m	ℓ_{max} & (ℓ_5), m	c
15	39.61	1.25 (1.25)	0	7.8	1.27 (1.25)	0	2.54	1.33 (1.27)	1
18	53.47	1.25 (1.25)	0	10.59	1.27 (1.25)	0	3.41	1.31 (1.27)	1
20	71.19	1.25 (1.25)	0	12.63	1.26 (1.25)	0	4.06	1.3 (1.27)	1

Table A.2. *S* = 0.25%

a) d_l = 10 mm

q litre/hr	d_m = 2 mm			d_m = 3 mm			d_m = 4 mm		
	H_T, m	ℓ_{max} & (ℓ_5), m	c	H_T, m	ℓ_{max} & (ℓ_5), m	c	H_T, m	ℓ_{max} & (ℓ_5), m	c
1	1.28	1.6 (1.43)	4	0.41	3.06 (2.18)	19	0.26	6.97 (4.19)	61
2	2.77	1.58 (1.42)	4	0.83	2.94 (2.11)	18	0.51	6.59 (3.99)	57
3	4.48	1.56 (1.41)	3	1.31	2.83 (2.06)	17	0.77	6.26 (3.82)	53
4	6.41	1.54 (1.40)	3	1.82	2.74 (2.01)	16	1.05	5.96 (3.67)	50
5	8.56	1.52 (1.39)	3	2.38	2.65 (1.97)	15	1.34	5.7 (3.54)	47
6	10.91	1.51 (1.38)	3	2.98	2.58 (1.93)	14	1.65	5.4 (3.42)	45
7	13.52	1.5 (1.37)	3	3.62	2.51 (1.90)	13	1.97	5.26 (3.31)	43
8	16.32	1.48 (1.37)	3	4.32	2.45 (1.87)	13	2.3	5.07 (3.21)	41
10	22.48	1.46 (1.36)	2	5.78	2.35 (1.81)	12	2.97	4.74 (3.04)	37
12	29.63	1.43 (1.35)	2	7.39	2.16 (1.77)	10	3.66	4.13 (2.90)	31
15	41.53	1.37 (1.34)	1	9.94	1.88 (1.71)	7	4.58	3.27 (2.72)	21
18	55.79	1.34 (1.33)	1	12.9	1.75 (1.67)	5	5.64	2.84 (2.58)	17
20	78.96	1.26 (1.25)	0	14.86	1.65 (1.57)	4	6.25	2.53 (2.27)	14

b) d_l = 12 mm

q litre/hr	d_m = 2 mm			d_m = 3 mm			d_m = 4 mm		
	H_T, m	ℓ_{max} & (ℓ_5), m	c	H_T, m	ℓ_{max} & (ℓ_5), m	c	H_T, m	ℓ_{max} & (ℓ_5), m	c
1	1.27	1.57 (1.42)	3	0.39	2.88 (2.12)	17	0.24	6.41 (4.00)	55
2	2.74	1.55 (1.41)	3	0.8	2.77 (2.06)	16	0.48	6.07 (3.82)	51
3	4.43	1.53 (1.40)	3	1.26	2.68 (2.01)	15	0.72	5.77 (3.66)	48
4	6.34	1.51 (1.39)	3	1.76	2.59 (1.96)	14	0.99	5.51 (3.52)	45
5	8.47	1.5 (1.38)	3	2.3	2.52 (1.92)	14	1.26	5.27 (3.39)	43
6	10.84	1.48 (1.37)	3	2.88	2.45 (1.89)	13	1.55	5.06 (3.28)	40
7	13.38	1.47 (1.37)	2	3.52	2.39 (1.86)	12	1.85	4.87 (3.18)	38
8	16.16	1.46 (1.36)	2	4.19	2.34 (1.83)	12	2.17	4.7 (3.09)	37
10	22.4	1.44 (1.35)	2	5.66	2.24 (1.78)	11	2.84	4.4 (2.93)	33
12	29.48	1.43 (1.34)	2	7.27	2.16 (1.73)	10	3.53	4.15 (2.79)	31
15	41.49	1.37 (1.33)	1	9.92	1.9 (1.68)	7	4.54	3.3 (2.63)	22
18	55.59	1.35 (1.32)	1	12.8	1.76 (1.64)	5	5.61	2.87 (2.49)	17
20	76.32	1.26 (1.25)	0	14.88	1.66 (1.62)	4	6.25	2.55 (2.42)	14

c) d_l = 14 mm

q litre/hr	d_m = 2 mm			d_m = 3 mm			d_m = 4 mm		
	H_T, m	ℓ_{max} & (ℓ_5), m	c	H_T, m	ℓ_{max} & (ℓ_5), m	c	H_T, m	ℓ_{max} & (ℓ_5), m	c
1	1.26	1.55 (1.41)	3	0.38	2.81 (2.09)	17	0.23	6.18 (3.92)	52
2	2.73	1.53 (1.40)	3	0.79	2.7 (2.04)	15	0.46	5.85 (3.74)	49
3	4.41	1.51 (1.39)	3	1.24	2.61 (1.99)	15	0.7	5.56 (3.59)	46
4	6.31	1.49 (1.38)	3	1.73	2.53 (1.94)	14	0.96	5.31 (3.45)	43
5	8.45	1.48 (1.38)	3	2.26	2.46 (1.90)	13	1.23	5.08 (3.33)	41
6	10.79	1.47 (1.37)	2	2.85	2.4 (1.87)	12	1.51	4.88 (3.22)	39
7	13.35	1.46 (1.36)	2	3.47	2.34 (1.84)	12	1.81	4.7 (3.12)	37
8	16.09	1.45 (1.36)	2	4.13	2.29 (1.81)	11	2.11	4.54 (3.03)	35
10	22.35	1.43 (1.35)	2	5.58	2.2 (1.76)	10	2.78	4.25 (2.88)	32

Table (Continued)

q litre/hr	d_m = 2 mm			d_m = 3 mm			d_m = 4 mm		
	H_T, m	ℓ_{max} & (ℓ_5), m	c	H_T, m	ℓ_{max} & (ℓ_5), m	c	H_T, m	ℓ_{max} & (ℓ_5), m	c
12	29.45	1.42 (1.34)	2	7.21	2.12 (1.72)	9	3.49	4.01 (2.75)	29
15	41.76	1.4 (1.33)	2	9.96	2.03 (1.67)	8	4.62	3.72 (2.59)	26
18	55.61	1.36 (1.32)	1	12.92	1.8 (1.63)	6	5.7	3.01 (2.46)	19
20	73.78	1.25 (1.25)	0	15.08	1.7 (1.61)	5	6.4	2.69 (2.38)	15

d) d_l = 16 mm

q litre/hr	d_m = 2 mm			d_m = 3 mm			d_m = 4 mm		
	H_T, m	ℓ_{max} & (ℓ_5), m	c	H_T, m	ℓ_{max} & (ℓ_5), m	c	H_T, m	ℓ_{max} & (ℓ_5), m	c
1	1.26	1.55 (1.41)	3	0.38	2.77 (2.08)	16	0.23	6.05 (3.88)	51
2	2.72	1.53 (1.40)	3	0.78	2.67 (2.02)	15	0.46	5.74 (3.71)	48
3	4.4	1.51 (1.39)	3	1.23	2.58 (1.98)	14	0.69	5.46 (3.56)	45
4	6.3	1.49 (1.38)	3	1.71	2.5 (1.93)	13	0.94	5.21 (3.42)	42
5	8.42	1.48 (1.37)	2	2.25	2.43 (1.90)	13	1.21	4.99 (3.30)	40
6	10.77	1.47 (1.37)	2	2.83	2.37 (1.86)	12	1.49	4.80 (3.19)	38
7	13.28	1.46 (1.36)	2	3.44	2.31 (1.83)	11	1.79	4.58 (3.10)	36
8	16.07	1.45 (1.36)	2	4.11	2.26 (1.80)	11	2.09	4.46 (3.01)	34
10	22.33	1.43 (1.35)	2	5.56	2.17 (1.75)	10	2.74	4.18 (2.85)	31
12	29.43	1.41 (1.34)	2	7.19	2.1 (1.71)	9	3.45	3.95 (2.73)	29
15	41.75	1.4 (1.33)	2	9.95	2.01 (1.66)	8	4.63	3.66 (2.57)	26
18	55.64	1.37 (1.32)	1	13.0	1.86 (1.62)	7	5.79	3.10 (2.44)	21
20	71.28	1.25 (1.25)	0	15.0	1.76 (1.6)	5	6.5	2.86 (2.37)	17

Table A.3. S = 0.5%

a) d_l = 10 mm

q litre/hr	d_m = 2 mm			d_m = 3 mm			d_m = 4 mm		
	H_T, m	ℓ_{max} & (ℓ_5), m	c	H_T, m	ℓ_{max} & (ℓ_5), m	c	H_T, m	ℓ_{max} & (ℓ_5), m	c
1	1.45	1.89 (1.59)	7	0.57	4.53 (2.99)	35	0.42	11.62 (6.77)	110
2	3.07	1.85 (1.57)	6	1.14	4.31 (2.88)	33	0.81	10.93 (6.40)	103
3	4.92	1.81 (1.55)	6	1.74	4.12 (2.78)	31	1.21	10.33 (6.08)	96
4	6.98	1.78 (1.53)	6	2.4	3.95 (2.69)	29	1.63	9.8 (5.80)	91
5	9.26	1.75 (1.51)	5	3.09	3.8 (2.61)	27	2.05	9.32 (5.55)	86
6	11.78	1.72 (1.50)	5	3.84	3.67 (2.54)	26	2.49	8.9 (5.32)	81
7	14.49	1.7 (1.49)	5	4.61	3.55 (2.47)	24	2.95	8.52 (5.12)	77
8	17.42	1.68 (1.48)	5	5.43	3.44 (2.41)	23	3.42	8.17 (4.94)	74
10	23.88	1.64 (1.46)	4	7.16	3.25 (2.31)	21	4.34	7.57 (4.62)	67
12	31.14	1.57 (1.44)	3	8.98	2.9 (2.23)	18	5.24	6.47 (4.35)	56
15	43.16	1.47 (1.42)	2	11.65	2.39 (2.12)	12	6.3	4.87 (4.01)	38
18	57.75	1.42 (1.40)	2	14.71	2.14 (2.04)	9	7.49	4.06 (3.75)	30
20	79.06	1.27 (1.26)	0	16.63	1.95 (1.85	7	8.04	3.47 (3.16)	24

b) d_l = 12 mm

q litre/hr	d_m = 2 mm			d_m = 3 mm			d_m = 4 mm		
	H_T, m	ℓ_{max} & (ℓ_5), m	c	H_T, m	ℓ_{max} & (ℓ_5), m	c	H_T, m	ℓ_{max} & (ℓ_5), m	c
1	1.43	1.86 (1.58)	7	0.55	4.35 (2.94)	33	0.4	11.07 (6.59)	104
2	3.04	1.82 (1.56)	6	1.1	4.15 (2.82)	31	0.78	10.41 (6.23)	97
3	4.87	1.78 (1.54)	6	1.69	3.97 (2.72)	29	1.16	9.84 (5.92)	91
4	6.92	1.75 (1.52)	5	2.33	3.81 (2.64)	27	1.56	9.34 (5.65)	86
5	9.19	1.72 (1.50)	5	3.01	3.66 (2.56)	26	1.97	8.89 (5.40)	81
6	11.69	1.7 (1.49)	5	3.74	3.54 (2.56)	24	2.4	8.49 (5.19)	77
7	14.36	1.68 (1.48)	5	4.5	3.42 (2.49)	23	2.84	8.13 (4.99)	73
8	17.26	1.65 (1.47)	4	5.31	3.32 (2.43)	22	3.29	7.8 (4.81)	70
10	23.81	1.62 (1.45)	4	7.04	3.14 (2.37)	20	4.25	7.23 (4.50)	64

Table A.3. (Continued)

q litre/hr	d_m = 2 mm			d_m = 3 mm			d_m = 4 mm		
	H_T, m	ℓ_{max} & (ℓ_5), m	c	H_T, m	ℓ_{max} & (ℓ_5), m	c	H_T, m	ℓ_{max} & (ℓ_5), m	c
12	31.16	1.59 (1.43)	4	8.92	2.99 (2.28)	19	5.18	6.76 (4.24)	59
15	43.64	1.49 (1.41)	3	11.74	2.48 (2.19)	13	6.39	5.15 (3.92)	41
18	57.8	1.44 (1.40)	2	14.87	2.22 (2.09)	10	7.67	4.31 (3.66)	33
20	76.41	1.26 (1.25)	0	16.92	2.02 (1.96)	8	8.31	3.7 (3.52)	26

c) d_l = 14 mm

q litre/hr	d_m = 2 mm			d_m = 3 mm			d_m = 4 mm		
	H_T, m	ℓ_{max} & (ℓ_5), m	c	H_T, m	ℓ_{max} & (ℓ_5), m	c	H_T, m	ℓ_{max} & (ℓ_5), m	c
1	1.43	1.84 (1.57)	6	0.54	4.28 (2.91)	32	0.4	10.83 (6.51)	102
2	3.03	1.8 (1.55)	6	1.09	4.08 (2.80)	30	0.76	10.19 (6.16)	95
3	4.85	1.77 (1.53)	6	1.67	3.9 (2.70)	28	1.14	9.64 (5.85)	89
4	6.89	1.74 (1.52)	5	2.31	3.74 (2.62)	27	1.53	9.14 (5.58)	84
5	9.16	1.71 (1.50)	5	2.98	3.61 (2.54)	25	1.94	8.71 (5.34)	79
6	11.66	1.69 (1.49)	5	3.69	3.48 (2.47)	24	2.36	8.32 5.13)	75
7	14.28	1.66 (1.48)	4	4.45	3.37 (2.41)	23	2.79	7.96 (4.93)	71
8	17.22	1.64 (1.46)	4	5.25	3.27 (2.36)	21	3.24	7.64 (4.76)	68
10	23.72	1.61 (1.45)	4	6.98	3.09 (2.26)	20	4.17	7.09 (4.45)	62
12	31.15	1.58 (1.43)	4	8.9	2.95 (2.18)	18	5.16	6.62 (4.2)	57
15	43.65	1.55 (1.41)	3	12	2.76 (2.08)	16	6.67	6.05 (3.88)	51
18	57.9	1.46 (1.39)	2	15.15	2.33 (2.00)	12	7.94	4.68 (3.63)	36
20	73.87	1.26 (1.25)	0	17.26	2.13 (1.95)	9	8.69	4.04 (3.48)	30

d) d_l = 16 mm

q litre/hr	d_m = 2 mm			d_m = 3 mm			d_m = 4 mm		
	H_T, m	ℓ_{max} & (ℓ_5), m	c	H_T, m	ℓ_{max} & (ℓ_5), m	c	H_T, m	ℓ_{max} & (ℓ_5), m	c
1	1.42	1.84 (1.57)	6	0.54	4.24 (2.90)	32	0.39	10.71 (6.47)	101
2	3.02	1.8 (1.55)	6	1.08	4.07 (2.79)	30	0.75	10.09 (6.12)	94

q litre/hr	d_m = 2 mm			d_m = 3 mm			d_m = 4 mm		
	H_T, m	ℓ_{max} & (ℓ_5), m	c	H_T, m	ℓ_{max} & (ℓ_5), m	c	H_T, m	ℓ_{max} & (ℓ_5), m	c
3	4.84	1.76 (1.53)	6	1.66	3.87 (2.69)	28	1.13	9.54 (5.82)	88
4	6.88	1.73 (1.51)	5	2.29	3.71 (2.61)	26	1.52	9.05 (5.55)	83
5	9.13	1.71 (1.50)	5	2.96	3.58 (2.53)	25	1.92	8.62 (5.31)	78
6	11.64	1.68 (1.49)	5	3.67	3.46 (2.46)	23	2.34	8.23 (5.10)	74
7	14.26	1.66 (1.47)	4	4.42	3.35 (2.40)	22	2.76	7.88 (4.91)	70
8	17.2	1.64 (1.46)	4	5.22	3.25 (2.35)	21	3.21	7.57 (4.73)	67
10	23.69	1.61 (1.44)	4	6.95	3.07 (2.25)	19	4.13	7.02 (4.43)	61
12	31.12	1.58 (1.43)	4	8.87	2.93 (2.17)	18	5.11	6.56 (4.18)	56
15	43.77	1.54 (1.41)	3	12.03	2.75 (2.07)	16	6.69	5.99 (3.86)	50
18	58.48	1.48 (1.39)	3	15.37	2.46 (1.99)	13	8.12	5.08 (3.61)	41
20	71.36	1.26 (1.25)	0	17.58	2.25 (1.95)	11	9.00	4.42 (3.47)	34

REFERENCES

Al-Amoud, A. I. 1995. Significance of Energy Losses Due to Emitter Connections in Trickle Irrigation Lines. *Journal of Agricultural Engineering Research,* 60, 1-5.

Allen, R. G., Pereira, L. S., Dirk, R. & Smith, M. 2011. Crop evapotranspiration: Guidelines for computing crop water requirements. *FAO Irrigation and Drainage Paper No. 56,* 1998.

Anyoji, H. & Wu, I. P. 1987. Statistical Approach For Drip Lateral DESIGN. *Transactions of the American Society of Agricultural Engineers,* 30, 187-192.

Begarello V, Ferro V, Provenzano G, Pumo D. 1995. Experimental study on flow-resistance law for small-diameter plastic pipes. Journal of irrigation and Draianage Engineering 121(5):313-316.

Benami, A. & Ofen, A. 1984. Irrigation Engineering. Sprinkler, Trickle and Surface Irrigation. Principles, Design and Agricultural Practices. *Irrigation Engineering: Sprinkler, Trickle, Surface Irrigation. Principles, Design and Agricultural Practices*.

Bhatnagar, P. R. & Srivastava, R. C. 2003. Gravity-fed drip irrigation system for hilly terraces of the northwest Himalayas. *Irrigation Science,* 21, 151-157.

Bhuiyan, M. A., Mohsen, M.F.N. and Elmasri, M.Z 1990. Microtubes as an alternative to pressure compensating emitters in drip irrigation systems. *Hydrosoft,Computational Mechanics Publications,* 3.

Blasius H. 1913. Das Ahnlichkeitsgesetz bei reibungsvorgangen in flussigkeiten. Forschg. Arb. Ing. -Wes. No. 134, Berlin.

Bralts, V. F. & WU, I. P. 1979. Emitter flow variation and uniformity for drip irrigation. *St. Joseph Mich.: ASAE.*

Bucks DA, Mayers LE.1973. Trickle Irrigation-application uniformity from simple emitter. Transaction of ASAE 16(6): 1108-1111.

Bucks, D. A., Nakayama, F. S. & Gilbert, R. G. 1979. Trickle irrigation water quality and preventive maintenance. *Agricultural Water Management,* 2, 149-162.

Capra, A. & Scicolone, B. 1998. Water quality and distribution uniformity in drip/trickle irrigation systems. *Journal of Agricultural Engineering Research,* 70, 355-365.

Christiansen, J. E. 1942. *Irrigation by sprinkling. California Agricultural Experimental Station Bulletin 670.*

Churchil SW. 1977. Friction-factor equation spans all fluid-flow regimes. American Institute of Chemical Engineers Journal 23:91-92.

Colebrook CF, White CM. 1937. The reduction of carrying capacity of pipes with age. Journal of Institution of Civil Engineers London 10: 115-122.

De Almeida, C. D. G. C., Botrel, T. A. & Smith, R. J. 2009. Characterization of the microtube emitters used in a novel micro-sprinkler. *Irrigation Science,* 27, 209-214.

Ella, V. B., Reyes, M. R. & Yoder, R. 2009. Effect of hydraulic head and slope of water distribution uniformity of a low-cost drip irrigation system. *Applied Engineering in Agriculture,* 25, 349-356.

Gilbert, R. G., Nakayama, F. S., Bucks, D. A., French, O. F. & Adamson, K. C. 1981. Trickle irrigation: Emitter clogging and other flow problems. *Agricultural Water Management,* 3, 159-178.

Hathoot, H. M., Al-Amoud, A. I. & Mohammad, F. S. 1993. Analysis and design of trickle-irrigation laterals. *Journal of Irrigation and Drainage Engineering,* 119, 756-767.

Hensen, V. E., Israelson, O. W. & Stringham, G. E. 1980. *Irrigation Principals and Practices.*

Hezarjaribi, A., Dehghani, A. A., Meftah Height, M. & Kiani, A. 2008. Hydraulic performances of various trickle irrigation emitters. *Journal of Agronomy,* 7, 265-271.

Howell, T. A. & Hiler, E. A. 1974. Designing Trickle Irrigation LATERALS for Uniformity. *J. Irrig. and Drain. Div. - Proc. A.S.C.E.,* 100.

Ito, H. 1960. Pressure losses in smooth pipe bends *Transaction of the ASME,* Series D, 82.

Ito H 1960. Pressure losses in smooth pipe bends. transaction of the ASME (series D) 82(1): 131-140.

Karmeli, D. & Peri, G. 1977. Design Procedure for a Sprinkler Lateral. 1, 225-235.

Keller, J. & Bliesner, R. D. 1990a. *Trickle Irrigation Design.*

Keller, J. & Bliesner, R. D. 1990b. Sprinkle and Trickle Irrigation. *New York, N.Y.: Van Nostrand Reinhold.*

Keller, J. & Karmeli, D. 1974. Trickle Irrigation DESIGN Parameters. *Transactions of the American Society of Agricultural Engineers,* 17, 678-684.

Khatri, K. C., Wu, I. P., Gitlin, H. M. & Phillips, A. L. 1979. Hydraulics of Microtube Emitters. *ASCE J Irrig Drain Div,* 105, 163-173.

Myers, L. E. & Bucks, D. A. 1972. Uniform Irrigation with Low-Pressure Trickle Systems. *ASCE J Irrig Drain Div,* 98, 341-346.

Nikuradse J.1933. Stromungsgesetze in rauhen Rohren. Forschg. Arb. Ing.-Wes. No.361, 1-22.

Özekici, B. & Sneed, R. E. 1995. Manufacturing variation for various trickli irrigation on-line emitters. *Applied Engineering in Agriculture,* 11, 235-240.

Perold, R. P. 1977. Design Of Irrigation Pipe Laterals with Multiple Outlets. *ASCE J Irrig Drain Div,* 103, 179-195.

Polak P, S. R. 1998. The potential contribution of low cost irrigation system to the improvement of irrigation productivity in India. *India—water resources management sector review: report on irrigation sector.* World Bank, Washington, D.C. In co-operation with India.

Schlichting H.1968. Boundary Layers Theory. 6th edn. MGraw-Hil; New York.

Singh, A. K., Rahman, A., Sharma, S. P., Upadhyaya, A. & Sikka, A. K. 2009. Small holders' irrigation - Problems and options. *Water Resources Management,* 23, 289-302.

Solomon, K. 1979. Manufacturing Variation Of Trickle Emitters. *Transactions of the American Society of Agricultural Engineers,* 22, 1034-1038, 1043.

Solomon, K. & Keller, J. 1978. Trickle Irrigation Uniformity and Efficiency. *ASCE J Irrig Drain Div,* 104, 293-306.

Solomon, K. H. 1985. Global Uniformity of Trickle Irrigation SYSTEMS. *Transactions of the American Society of Agricultural Engineers,* 28, 1151-1158.

Tscheschke, P. D. 1973. Trickel irrigation salinity patterns as influenced by irrigation levels and application rates. *M.S. Thesis, Utah State University, Logan, Utah* 115p.

Vallesquino, P. 2008. An approach for simulating the hydraulic performance of irrigation laterals. *Irrigation Science,* 26, 475-486.

Vermeiren, I. & Jobling, J. A. 1980. *Localized Irrigation,* 36, 202.

von Bernuth, R. D. 1990. Simple and accurate friction loss equation for plastic pipe. *Journal of Irrigation and Drainage Engineering,* 116, 294-298.

Watters, G. Z. & Keller, J. 1978. Trickle Irrigation Tubing Hydraulics. *Paper - American Society of Agricultural Engineers.*

Wu, I. P. 1992. Energy gradient line approach for direct hydraulic calculation in drip irrigation design. *Irrigation Science,* 13, 21-29.

Wu, I. P. & Gitlin, H. M. 1973. Hydraulics and uniformity for drip irrigation. *J. Irrig. Drain Div., PROC. ASCE,* 99.

Wu, I. P. & Gitlin, H. M. 1977a. Design of drip irrigation lines with varying pipe sizes. *J. IRRIG. DRAIN. DIV.: PROC. ASCE.,* 103 , IR4, Dec.1977, 499-503.

Wu, I. P. & Gitlin, H. M. 1977b. Design of drip irrigation submain. *ASCE PROC.,* 103, 231-243.

Yildirim, G. 2007. Analytical relationships for designing multiple outlets pipelines. *Journal of Irrigation and Drainage Engineering,* 133, 140-154.

In: Drip Irrigation
Editor: Alfred H. Steele

ISBN: 978-1-63463-737-4

Chapter 5

GROWTH AND WATER USE EFFICIENCY OF A MULTI-VARIETAL DRIP IRRIGATED OLIVE (*OLEA EUROPAEA* L.) TREES CULTIVATED IN NORTHERN TUNISIA

Masmoudi-Charfi Chiraz * ***and Msallem Monji, Researchers***
Ben Abdallah Sirine, Technical Collaborator
Olive Tree Institute, Tunisia

ABSTRACT

This work was carried out in northern Tunisia (36.5ºN, 10.2ºE) in order to study the behavior of local and foreign olive varieties (Olea europaea L.) cultivated in the same area under drip irrigation. Through this experiment we entend to evaluate the ability of these cultivars (collection) to valorize water. For this purpose, the water use efficiency was determined for both fruit growth ($WUE_{fr,}$ mm of fruit increase / m^3 I+P) and olive production (kg of fruits / m^3 I+P). During the three years of monitoring, 2010-2012 and 2013, trees were irrigated from May to September twice a week. Each tree received an amount of water equal to 5.7 m^3 in 2010 and 3.4 m^3 in 2012 and 2013. Results showed that fruits grew continuously from fruit set to harvest with different trends following the season and variety even during pit hardening. Four models of fruit growth were identified, with one or two peak values depending on cultivars. Most growth (47%) occurred during the early stage of fruit development (Stage 1). This resulted in high WUE_{fr}, of 12.1 mm/m^3 in 2010, 9.6 mm/m^3 in 2012 and 6.9 mm/m^3 in 2013. During stage 2, fruit growth slowed and some diameter shrinkage was recorded for cvs., Koroneiki (-1.4%), Branquita (-7.7%), Sayali (-20.7%) and Lucques (-22.1%) due to water shortage. Some varieties showed high rates of growth during stage 3. For the overall growing cycle, the average water use efficiencies reached 1.0 mm/m^3 in 2010, 4.2 mm/m^3 in 2012 and 3.3 mm/m^3 in 2013, with important differences between cultivars (0.20-6.22 mm/m^3). Maximums of up to 2 mm/m^3 were obtained for cvs., Barouni and Madurel, and the minimums were observed for cvs., Koroneiki, Arbequina, Chemlali and Coratina. Olive production varied consistenly between varieties and trees of the same cv.,. Average WUE_p were equal to 1.49 kg/m^3 in 2010, 2.02 kg/m^3 in 2012 and 3.28 kg/m^3 in 2013, with maximums observed for cvs., Chemlali (2.11 kg/m^3 in 2010 and 10.06 kg/m^3 in 2012) and Calegua (7.74 kg/m^3). The lowest values of

* E-mail: masmoudi.chiraz@yahoo.fr.

water use efficiencies for olive production were recorded for cvs., Arbequina in 2010 (0.72 kg/m^3), Manzanille in 2012 (0.03 kg/m^3) and Sayali in 2013 (0.59 kg/m^3). To conclude, we can say that most varieties valorize water more efficiently during their early fruit growth process, i.e., just after fruit set, with however some exceptions and large disparities between cultivars. This should be taken into account when planting multi-varietal olive orchards in order to get harmonious multi-varietal plantations. In such 'orchards', particular attention should be given to tree vigor and soil coverage, which was found to be correlated to WUE and olive production. Values of WUE for fruit growth and olive production presented herein can be used indicatively for this purpose depending on the objective of planting such olive groves: WUE_{fr} is usefull for table olive orchards, while WUE_p is more suitable for oil production ones.

Keywords: Soil coverage, soil water content, fruit growth, WUE for fruit growth and olive production

1. Introduction

In Tunisia most olive orchards are cultivated under rainfed conditions with annual rainfall amounts ranging between 80 mm (south) and 500 mm (north). Precipitations vary consistenly with time and site of olive cultivation, making growth and production of olive trees largely dependent on the amount of water received during the fruit growth period; the annual national production ranges between 60000 and 300000 tones of olives depending on the climatic conditions that prevail during the actual year and the previous one (DGPA, 2006).

In the central and southern areas, yields rarely exceed 0.7 ton ha^{-1}, resulting in low fruit size and water use efficiency (0.6-1.0 kg of olives/m^3). There, irrigation is applied with more or less regular inputs as a compulsory practice only when high-fruit-production is expected to prevent fruit drops (June-September). In the north of the country, maximum yield reaches 3.0 tons ha^{-1} for an annual input of 500 mm of water, including effective rainfall. Water is supplied in most cases by surface irrigation, resulting in important loses of water and low water use efficiency, not exceeding 50%. This problem is not specific to olive; it is a major constraint facing irrigated crops in general; only 20% of orchadists practice drip irrigation with average efficiency of about 60-70%. Regular drip irrigation is applied in high tree density olive plantations of up to 200 trees/ha in order to improve yield, stabilize production and reduce alternate bearing. However, when prospecting olive orchards by a specialised group of Tunisian researchers, it appears that the applied amounts of water, often, don't match the crop water needs. The technical and scientific papers published by Masmoudi-Charfi et al., (2004), Masmoudi-Charfi, (2006), Masmoudi-Charfi et al., (2012a and 2012b), and Masmoudi-Charfi and Habaieb, (2014), provide valuable estimates of water requirements of olive groves and the specific irrigation amounts following the area of olive cultivation, age of trees, soil coverage and their stage of development. The Penman-Monteith method was used for this purpose (Allen et al., 1998). In these papers, informations are provided to help growers to manage more efficiently their orchards and to save more water, which should be applied at reasonable amounts and at specific periods of the growing cycle.

Indeed, researches made around the Mediterranean basin, dealing with water use and irrigation of olive orchards showed that olive trees respond favourably and quickly to water application even when low volumes are applied (Wullschleger et al., 1998; Tognetti et al.,

2005; Ben Ahmed et al., 2007), but the response to water vary substantially depending on variety, the amount of water received and timing (Palomo et al., 2002; Pastor 2005; Orgaz et al., 2006; Testi et al., 2006; Grattan et al., 2006; Melgar et al., 2008; Chehab, 2009; Iniesta et al., 2009; Aïachi et al., 2009 and 2012a; Masmoudi-Charfi et al., 2010; Martin-Vertedor, 2011b; Bamouh et al., 2012). Water applied at the beginning of the growing cycle improves shoot length and increase the potential sites for olive production (Michelakis and Vougioucalou, 1988; Michelakis, 1995 and 2000; Tognetti et al., 2006; Aïachi Mezghani, 2007; Perez-Lopez et al., 2007; Palease et al., 2010), while irrigation practiced later, during the fruit enlargement stage affects fruit size at harvest, trunk diameter increases and oil production (Michelakis, 1990 and 2000; Fernandez et al., 2003; Moriana et al., 2003; Connor and Fereres, 2005; Grattan et al., 2006; Melgar et al., 2008; Iniesta et al., 2009; Palese et al., 2010). Under drip irrigation, Ben Ahmed et al., (2007), Chehab et al., (2009) and Aïachi et al., (2009, 2012a, 2012b) observed consistent increases of yields by 3 to 5 times in comparaison to the control depending on cultivar and the watering and growing conditions, but in all cases, ensuring a gain of water of about 40%. This result is very important because fresh water becomes scarcer in Tunisia (Hamza, 2009). Economic strategies applied for peaches and some olive plantations, i.e., *Regulated Defict Irrigation and Partial Root Drying* should be generalized for olive (Goldhamer, 1999; Fernandez et al., 2003; Moriana, 2003; Wahbi et al., 2005; Fernandez, 2006; Tognetti et al., 2006; Fernandez et al., 2006; Ghrab et al., 2013). Experiments made in Central Tunisia on cvs., Chétoui, Manzanilla, Picholine and Coratina with amounts of water ranging between 20% ET_c and 100%ET_c (Masmoudi-Charfi et al., 2010) showed that water applied regularly, twice a week at 20%ET_c (ET_c=582 mm), from fruit set to early September, provided hight water use efficiencies (WUE) ranging between 1.6 kg m^{-3} and 3.9 kg m^{-3}. For the Chemlali cultivar, which is the main Tunisiain variety, maximum WUE was 1.2 kg m^{-3}obtained at 50% ET_c. The lowest WUE is recorded at 100%ET_c. Furthermore, supplying trees with suitable amounts of water, given at precise stages of tree and fruit development, provide a better control of nutrients and carbohydrate's distribution, leading to more regular yields (Androulakis, 1997; Fernandez and Moreno, 1999; D'Andria et al., 2008; Iniesta et al., 2009; Palease et al., 2010; Bchir et al., 2013).

However, although irrigation is considered as the major factor controlling olive production and fruit size, other factors should be taken into account like climate, variety, fertilization, pruning, fruit load, competition for nutrients …etc. (Forshey et al., 1989; Suarez et al., 1984; Proietti and Tombesi, 1996; Patumi et al., 2002; Gucci et al., 2007). Indeed, these conditions can reduce substantially yields even if irrigation is applied correctly. It is known that high temperatures exceeding 35°C can stop growth and eliminate the crop because they are often associated to water stress and high light fluence (Bonji and Palliotti, 1994). Also, the influence of the training system on photosynthesis and productivity is well known. Positive effect of high LAI on fruit growth, size of olives and production is reported in many papers (Proeitti and Tombesi, 1996; Fernandez and Moreno, 1999). Excurrent shapes of trees provide higher values of LAI (4.2), while the decurrent forms give lower values of about 2.5. Fertilisation is also important to consider because most of the Tunisian soils are poor and ferti-irrigation is not widespread. However, the choice of cultivars remains the most important factor to consider when planting olive (Olea europae, L.) orchards because trees takes many years to reach full size, to come into full bearing and living long time. This choice depends on the cultivation system which would be adopted and it is obviously imposed by the aim of installing olive plantations. Therefore, we should verify if the variety can be suitable and well adapted to the environmental conditions, taking into account the amount of available water and the tolerance of the variety to aridity and water deficit, what kind of horticultural package should be applied that allows good development of the orchard, optimal production, large benefits and longer economical live of the tree. In the new cultivation olive orchards, reduction of tree vigor is taken into consideration via selection of adapted cvs., allowing mechanization of harvest and an effective use of water (Larbi, 2009). In the intensive orchards, panoply of cultivars is used. The cvs., Meski, Chétoui and Chemlali dominate the Tunisian landscape and are in many cases associated to foreign cvs., like Manzanille, Koroneiki and Picholine. Other local varieties like Chemchali, Oueslati (Trigui, Msallem and Co, 2002) having high productivity and good oil quality are valorized in their area of origin.

In this study, carried out in east-north of Tunisia, we aim to compare the behavior of panoply of local and foreign olive varieties growing in the same area, under intensive conditions and drip irrigation. Cultivars were classified following their capacity to valorize water for fruit growth and olive production through determination of their water use efficiencies.

2. Materials and Methods

2.1. Site of Experiment and Climatic Conditions

A field experiment was carried out from 2010 to 2013 (except 2011) in northern Tunisia on panoply of olive varieties cultivated at the *Experimental Research Station of the National Institute of Researchers in Rural Engeneering, Water and Forests,* located at Nabeul, about 60 km from the capital Tunis. The region, named 'Cap-Bon' is a coastal and touristic, producing vegetables and strawberries of high quality.

In this region, olive is cultivated under irrigation at densities of up to 200 trees/ha. Climate is semi-arid (Index of Aridity according to the Thornwaite calculation procedure is 0.29, Masmoudi-Charfi and Habaieb, 2014) with annual rainfall of about 350 mm and reference evapotranspiration (ET_0) approximating 1300 mm. It is being dry and hot from May to October and rainy in autumn. Analyses of rainfall series showed large variations between years and months with minimum values recorded in July and maximums in December. Daily reference evapotranspiration (ET_0) was determined following the Penman-Monteith equation (Allen et al., 1998). An average value of ET_0 was used to establish the irrigation program. All weather variables were obtained from the *National Institute of Meteorology (INM).*

The orchard was planted in 2003 at 278 trees/ha by using own rooted olive cuttings of local and foreign cultivars. The experimental set-up was organized as shown in figure 1, considering three olive trees per variety. Results were provided only for 29 olive cvs., excluding younger trees and varieties that have missing trees.

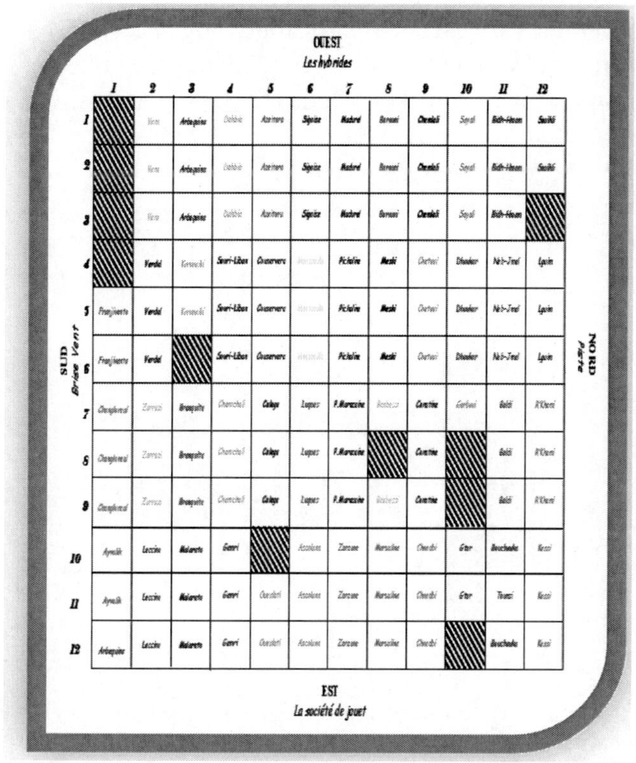

Figure 1. Experimental design of the olive orchard. Nabeul-Tunisia. The experiment involved panoply of cultivars cultivated under drip irrigation.

The soil of the orchard is sandy, non-calcareous and poor in organic matter. Crop management practices *i.e.,* pruning, fertilization, weed control and pest control were made following the standards but taken at minimum intervention (Masmoudi-Charfi et al., 2006; Masmoudi-Charfi et al., 2012b). Trees were pruned every two years and trained following the monocone shape. The orchard was plowed 3-4 times a year. Drip irrigation was practiced during the summer months, beginning from half of May with fresh water provided by wells, and containing an amount of salts ranging between 2.5 and 3.5 g/L depending on the season. Two lines of emitters were used. Drippers were placed at 0.33 m each one from the other and have a nominal discharge of 4 L/hour.

2.2. Crop Water Needs and Irrigation Application

Crop water needs (ET_c, mm) were determined following the FAO climatic method as $ET_c = ET_0 \, x \, K_c \, x \, K_r$; monthly reference evapotranspiration (ET_0) values were computed according to the Penman-Monteith equation (Allen et al., 1998), K_c was taken equal to 0.5 for eight to ten years old trees and K_r was determined on the basis of the tree downward projection canopy flat area (Masmoudi-Charfi et al., 2004). For this purpose, canopy diameters were measured on all trees at the beginning of the irrigation period and soil coverage was determined for each tree separately (Figure 2).

Values of soil coverage varied consistently following the variety and years'conditions. Low values of less than 15% were recorded for cvs., Franjivento (6.3%), Changloreal (9.7%), Branquita (11.8%), Souri Liban (8.4%), Gemri (7.7%), Azeitera (9.3%) and Besbassi (12.4%). High percentages of up to 30% were observed for cvs., Coratina (50.1%), Chemlali (39.5%), Zarzane (37.5%), Madurel (32.7%) and Malarato (46.6%).

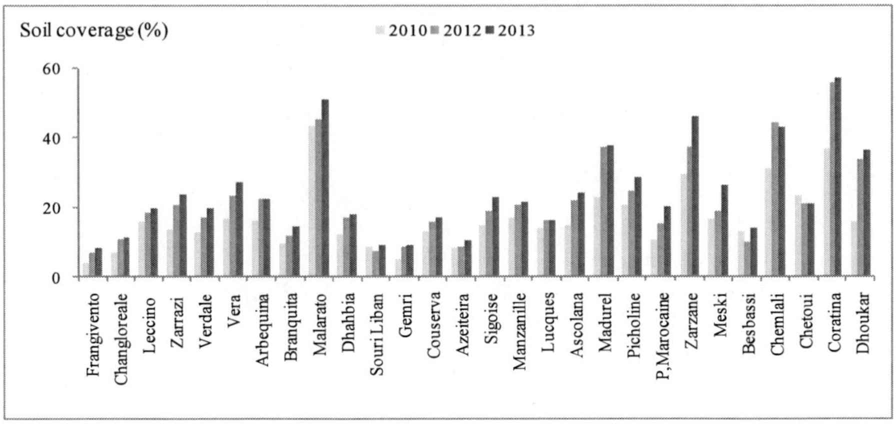

Figure 2. Soil coverage (%) recorded during the three growing experimental years following variety. Multi-varietal Olive orchard of Nabeul.

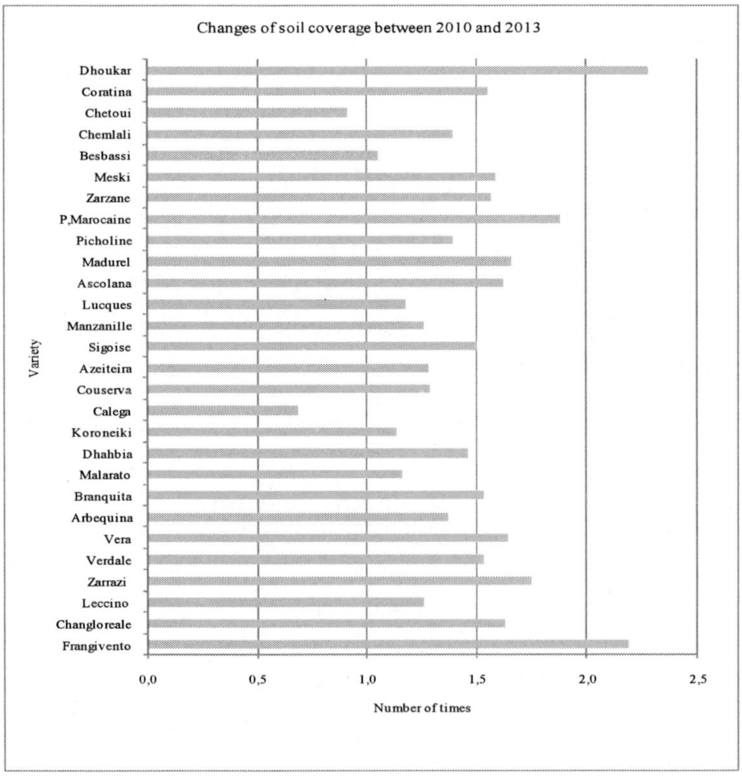

Figure 3. Increase of soil coverage between 2010 and 2013 following variety. Multi-varietal Olive orchard of Nabeul.

The consistent changes of soil coverage recorded between 2010 and 2013 are due to pruning and shoot development. Highest increases were observed for Doukhar (x2.3), Franjivento (x2.2) and Picholine Marocaine (x1.9) as shown in figure 3. The lowest values were recorded for cvs., Chétoui (x0.9), Souri Liban and Besessi (x1.1). Average increase was equal to 1.5. Water requirements were determined on the basis of the average percentages, equal to 16.7% in 2010, 21.8% in 2012 and 24.2% in 2013.

Reference evapotranspiration (ET_0-PM) computed for 10-days period are reported in Table 1 with seasonal value of 713 mm recorded from May to September. The crop water needs determined on the basis of these estimates and the average soil coverage values, varied between 38.4 mm (September) and 57.4 mm (July). The peak value was reached during the last decade of July. Seasonal requirements (May-September) are of 250 mm, equivalent to 8.9 m^3/ tree. Ratio of ET_c to ET_o is equal to 0.35 ($K_c = 0.7$ and $K_r = 0.5$).

Table 1. Reference evapotranspiration (ET_0, mm/decade), crop water needs (ET_c, mm/decade) and corresponding amounts of water needed each decade for eight to ten years old olive orchard

	Decade	ET_o	ET_c	60%ET_c
May	1 - 10	35.6	12.5	7.5
	11 - 20	33.7	11.8	7.1
	21 - 31	53.2	18.6	11.2
June	1 - 10	51.7	18.1	10.9
	11 - 20	51.1	17.9	10.7
	21 - 30	54.1	18.9	11.4
July	1 -10	50.7	17.7	10.6
	11 - 20	54.0	18.9	11.3
	21 - 31	58.2	20.4	12.2
August	1 - 10	52.9	18.5	11.1
	11 - 20	51.0	17.9	10.7
	21 - 31	57.2	20.0	12.0
September	1 - 10	41.5	14.5	8.7
	11 - 20	34.2	12.0	7.2
	21 - 30	33.9	11.9	7.1
Total		713	250	150

The program of irrigation was established initially to apply 60%ET_c (Table 1), but the lack of water recorded during those years and particularly during the summer periods did not allow regular application of the programmed amounts. In addition to the frequent failures of the irrigation system which was implemented many years ago, and does not allow irrigation of each tree individually. Irrigation was then applied equally for all trees. Amounts of water applied monthly are reported in figure 4.

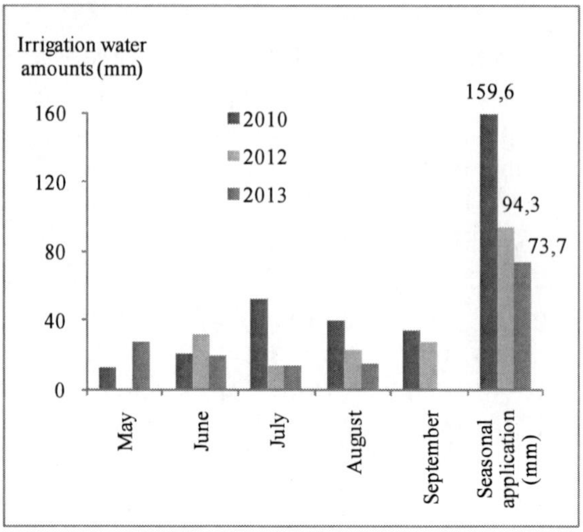

Figure 4. Monthly and seasonal irrigation water amounts (mm) applied to the olive orchard during the campaigns 2010, 2012 and 2013.

In 2010, irrigation was practiced from 5 May to 20 September, with amounts ranging between 12.8 mm and 52.1 mm/month and a seasonal amount of 159.6 mm, equivalent to 64%ET_c. Some water shortage was recorded from 25/5 to 10/6. In 2012, irrigation was practiced from June (1/6) to early September (4/9) with total amount of 94.3 mm and monthly values ranging between 13.2 mm and 31.5 mm. In 2013, trees were irrigated from mid May to 22th August and received a seasonal amount of 73.7 mm with monthly values varying from 13.2 mm to 26.8 mm. Each tree received an irrigation water amount of 5.7 m^3 in 2010, 3.39 m^3 in 2012 and 3.44 m^3 in 2013. During the irrigation period, soil water content was determined regularly below canopies (SF) considering regular measurements of soil moisture (H_v,%), made gravimetrically to 1 m depth. Values were regularly compared to those obtained within the non-watered area (at the intersection point of 4 neighbouring olive trees, I).

2.3. Fruit Growth Monitoring, Yield and Water Use Efficiency

Fruit growth was monitored regularly in situ with 1/100 calliper from fruit set to harvest, considering 5 fruits per variety. At harvest, fruit and endocarp length (L), width (D) and weight (P_f) were determined on the same fruits. Production of each tree was weighed individually; the average annual and cumulative yields were computed considering the three years of monitoring. Average water use efficiencies (WUE) were determined for fruit growth (mm/m^3 I+P) and olive production (kg/m^3 I+P). All efficiencies were reported to the amount of water received during the irrigation season (m^3 of P+I), *i.e.*, rainfall and irrigation amounts.

2.4. Analyses of Data

Average values were computed for each variable as well as ecartypes. Correlations between the different growth parameters were investigated particularly those concerning soil coverage in relation to olive production and water use efficiencies (WUE_{fr}, mm increase/m^3 and WUE_p, kg/m^3). Varieties were then classified following their capacity to valorize water on the basis of fruit size and water use efficiencies. Statistical analyses were performed through the establishment of linear regressions between fruit growth parameters, production, water use efficiencies and soil coverage.

3. RESULTS

3.1. Irrigation Management

Irrigation was practiced annually during 5 successive months depending on year's conditions with frequencies ranging between 3 and 21 days following the season. Table 2 presents the seasonal ratios between the crop water needs and the irrigation application amounts, providing information about the recovery of water requirements at seasonal and annual scales.

Rainfall amounts received during the irrigation period were 25 mm in 2010, 32 mm in 2012 and 20.9 mm in 2013. Accounting for an effective rainfall amount (P_e) of about 70%, the ratios $(P_e+I)/ET_c$ were equal to 71% in 2010 and 56% in 2012 and 2013. At annual scale the ratios reached 89%, 82% and 72% in 2010, 2012 and 2013, respectively.

Some water shortage was felt in 2010 and in 2012 at the beginning of the season, while a massive amount of water was applied during the ultimate stages of fruit development. This affected consistently fruit growth and the water use efficiency.

Table 2. Irrigation parameters relative to the multi-varietal orchard of Nabeul

Parameter	2010	2012	2013
Irrigation amount (mm/season)	159.6	94.3	73.7
Irrigation amount (m^3/ tree/ season)	5.7	3.39	3.44
Frequency (day)	3-20	4-20	8-21
Period of irrigation	5/5-20/9	1/6 - 4/9	15/5- 2/8
P+I (mm/year)	360	330.3	291.3
$P + I / ET_c$ (year)	0.89	0.82	0.72
P_e +I (mm/season)	177.1	116.7	88.3
$P_e + I / ET_c$ (season)	0.71	0.56	0.56

3.2. Soil Water Content

During the month of April, coinciding with the period of flowering, trees received only water provided by rainfall. Thus development of flowers was ensured under limited water amounts (Figure 5). During the following month (May), soil water content values recorded at the same sites were greater. This period is that of fruit set. Similar situation was recorded in June, when pits harden. It was the driest period. Low amounts of water were recorded in the non-watered area due to soil evaporation, which increased with the increasing climatic demand.

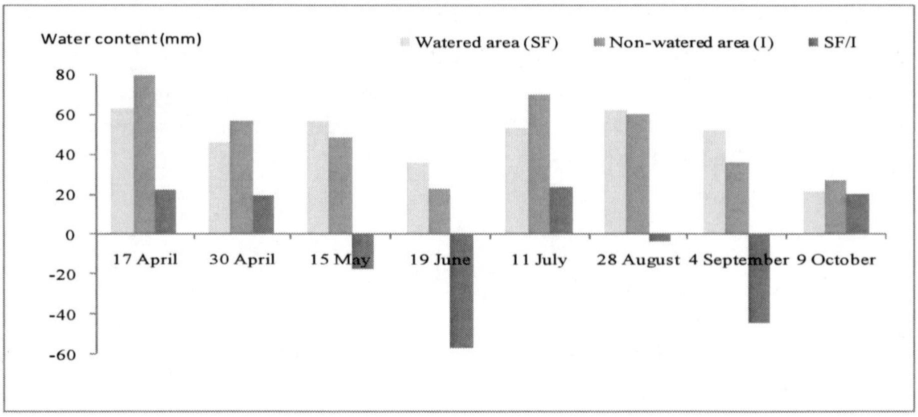

Figure 5. Soil water content (mm) determined during the campaign 2012 within the watered (SF) and non-watered (I) areas to 1 m depth.

Low water content values were observed in July due to the activity of the trees. Indeed, during this period, trees support fruit growth, synthesis of oil and development of new buds, which are induced during summer and continue their development over the following winter. In August and October, certain homogeneity was recorded between the watered and non-watered areas.

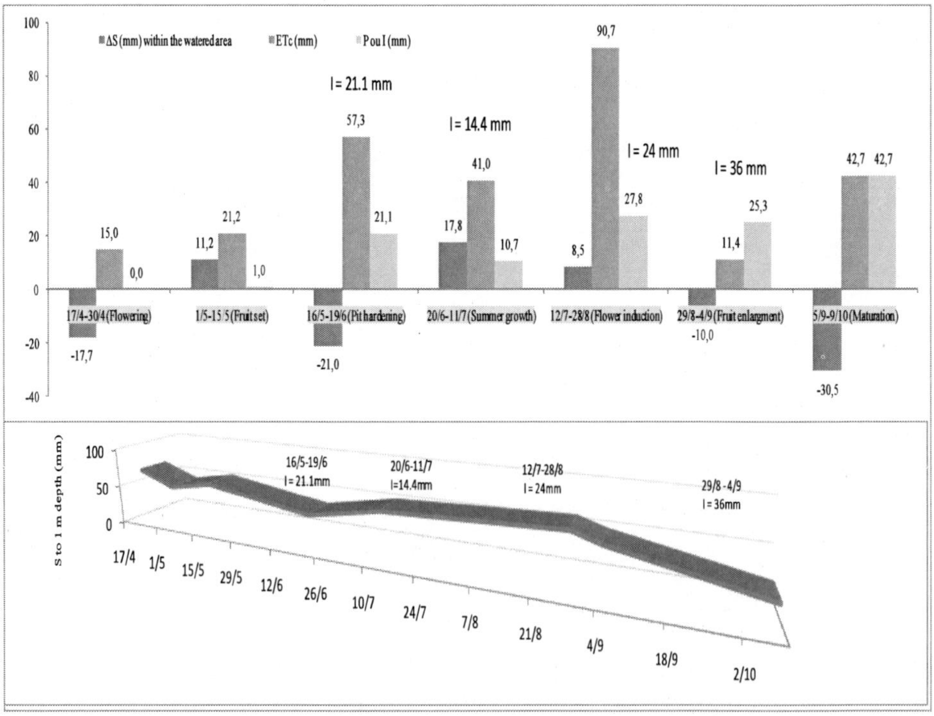

Figure 6. Irrigation parameters recorded during the campaign 2012. Stocks of water were determined gravimetrically between two successive measurements. Water requirements (ET_c, mm) and supplies (P and I, mm) are indicated in the same figure with regard to the growing periods of the olive trees.

During the irrigation season, soil water contents varied following depth. Maximum values were recorded at the surface and the minimums were observed within the down ground horizons. Differences between measurements were important during the period of pit hardening; lower values were recorded below the canopies. Superficial (20 cm) and deep (> 80 cm) horizons present peak values during the months of March and June. The medium horizons presented their peak values one month later between end of April and end of June (Figure 6).

Comparisons made between the zones concerned with water content monitoring on the basis of the ratio of I/SF showed:

0-20 cm: I/SF=1 late January, in April and September. During the fruit set period, I > SF.

20-40 cm: I/SF < 1, recorded in September.

40-60 cm: I/SF > 1 from January to end of April, characterized by intense fructification activity. During the irrigation period coinciding with the fruit growth period, I < SF.

60-80 cm: I/SF > 1.

80-100 cm, low values of I/SF were recorded between early April and end of June. Values exceed the unit in all cases.

Granulometric analyses showed differences in soil texture between the up ground and down ground horizons with high % of sand within the deeper layers, explaining the previous observations. Disparities were also due to the different processes involved in water depletion.

During the summer months, values of (P+I) – ET_c were negative ranging between -15.0 mm and -63.0 mm in 2012 and -32.4 mm and -78.2 mm in 2013. These watering conditions affected significantly fruit development and their water use efficiency.

3.3. Fruit Growth

3.3.1. Pomology of Fruits

Fruits grew from fruit set to harvest with different trends depending on the period of growth, year's conditions and variety. Early measurements of fruit diameter (D, mm) carried out during the 3-growing seasons, appromimately at the same period, *i.e.,* late of June (21/6 for 2010 and 26/6 for 2012 and 2013) showed important disparities between years and amongt varieties. The largest differences were recorded for the cultivars Branquita, Gemri and Sayali. At harvest, the average pomological records monitored on ripe and well colored fruits, reached 2.22 cm for fruit length (L), 1.70 cm for the equatorial diameter (D), 1.34 for L/D ratio, 4.01 g for fresh fruit weight (P_f, g) and 54.64% for water content. Maximum records reached 3.17 cm for L (Lucques), 2.55 cm for D (Souri Liban), 1.89 for L/D ratio (Lucques), 12.45 g for P_f (Zarrazi) and 64.59% for water content (Barouni). Minimum observations were of 1.31 cm for L (Souihli), 0.88 cm for D (Koroneiki), 1.08 for L/D (Ayvalik), 0.43 g for P_f (Koroneiki) and 42.03% for water content (Malarato). Stone's characteristics, which are considered as standars for a variety and used in genetic to characterize varieties, showed also large disparities between cultivars with average values of 14.30 mm for L, 7.35 mm for D and 0.51 g for P_f. Maximum stones'records were observed for cv., Besbassi, reaching 21 mm, for L, 10 mm for D and 1.37 g for P_f. Minimum values, recorded for cv., Souihli, were equal to 9.8 mm for L and 5 mm for D. The cv., Ayvalik presented the smaller pits with P_f of 0.1 g. Compared to the fruit fresh weight, that of pits represented in average 13.51% with maximum of 21% recorded for cvs., Dhoukar and Madurel and minimum of 8% recorded for Picholine.

3.3.2. Stages of Fruit Development

Fruit develop following three growing periods with different trends. The first period of fruit growth is rapid and short, occuring from full bloom to pit hardening, which ends during the second half of June. Rates reached 0.25 mm/day with large differences between cultivars. The second period is that of pit hardening and prolonged over a month approximately; it occured with lower rate, rarely exceeding 0.15 mm/day. During this period the embryo develop. The third period takes place beginning from end of September. Fruits color gradually and grew with low rates, of 0.1 mm/day and less, toward complete maturation.

Results showed some diameter shrinkage during stage 2 in 2010, affecting particularly the cvs., Koroneiki (-1.4%), Branquita (-7.7%), Sayali (-20.7%) and Lucques (-22.1%) due to water shortage. Similar behavior was recorded in 2013 for cvs., Dahbia (-3%) and Chemlali (-8%). Other cultivars stopped growing for about 20-days long period from end of June to 10 of July.

Increases of fruit diameter varied following cultivars. In 2013, fruit diameter gains reached, in average, 46% for stage 1, 17% for stage 2 and 37% for stage 3, with maximums recorded for Coratina (74%, Stage 1), Branquita (40%, Stage 2) and Rkhami (59, Stage 3), while minimum increases were observed for cvs., Rkhami (29%, Stage 1), Sayali (6%, Stage 2) and Coratina (1%, Stage 3). For most cultivars, more than 40% of fruit gain was recorded during stage 1. However, there were some exceptions concerning the cultivars Rkhami, Souihli, Manzanille and Franjivento, reaching during stage 1, 29%, 33%, 36% and 31% of their full size, respectively (Figure 7).

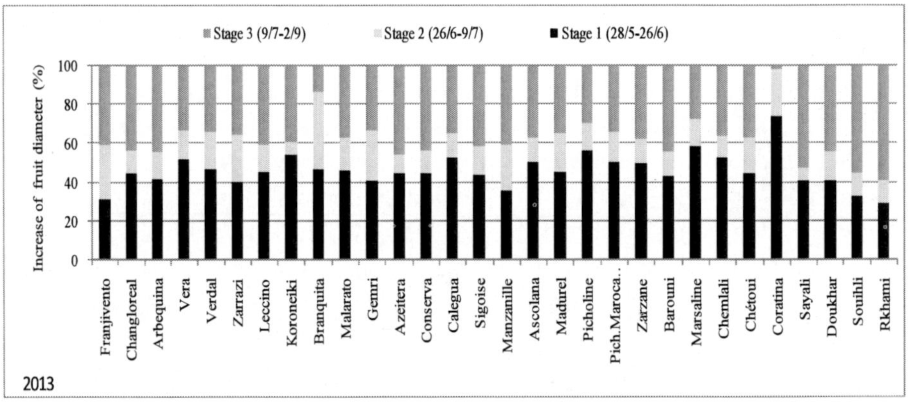

Figure 7. Increases of fruit diameter (%) following the stage of development and variety.

The cultivars Vera, Verdale, Zarrazi, Branquita, Gemri, Calegua, Madurel, Picholine, Picholine Marocaine, Marsaline and Coratina achieved more than 65% of fruit size before mid-July. The cultivars Branquita and Coratina attained more than 87% of their final volume at then end of stage 2. In opposite, varieties like Franjivento (69%), Zarrazi (60%), Manzanille (64%), Souihli (67%), Rkhami (71%) achievd their major growth process during stages 2 and 3, *i.e,* after pit hardening. These differences are very important, showing that olive varieties valorize water supplies differently. Some of them are more able to profit from water during their early fruit growth process, just after fruit set, while others valorize water more efficiently after pit hardening. Association of cvs., within the orchard should take into account this behavior. In addition to the fact that although cvs., evolved under the same environmental conditions, some of them showed regular increases of fruit diameter, while others presented some diameter shrinkage when water becomes less available, allowing comparaison and classification of these varieties.

3.3.3. Fruit Growth and Water Use Efficiency (WUE_{fr})

Water use efficiency is defined herein as the ratio of fruit increase to the amount of water supplied by irrigation and rainfall (WUE_{fr}, mm/m^3 I+P).

Seasonal WUE_{fr} recorded from full bloom to harvest varied consistently between years and among varieties (Figure 8). Results showed lowest values in 2010 ranging between 2.34 and 0.20 mm/m^3 I+P for a seasonal water supply of 5.7 m^3/tree. The average efficiency WUE_{fr} approximates 1.0 mm/m^3 I+P. Maximum values were observed for cvs., Barouni (2.34 mm/m^3 I+P) and Madurel (2.2 mm/m^3 I+P) and the minimum was that of Koroneiki. In 2012, values of WUE_{fr} ranged between 6.22 and 2.33 mm/m^3 I+P with an average efficiency of 4.2 mm/m^3 I+P determined for a seasonal water application of 3.41 m^3/tree, representing 60% of the amount applied in 2010. The following year, 2013, lower efficiencies were recorded ranging between 4.45 and 1.99 mm/m^3 I+P although the amount of water applied was similar (3.39 m^3/tree). Average efficiencies for fruit growth were determined for all cvs., except Meski, Lucque, Souri Liban, Ayvalik, Gtar, Beldi, Dhahbia, Chemchali, Besbassi and Tounsi due to lack of data.

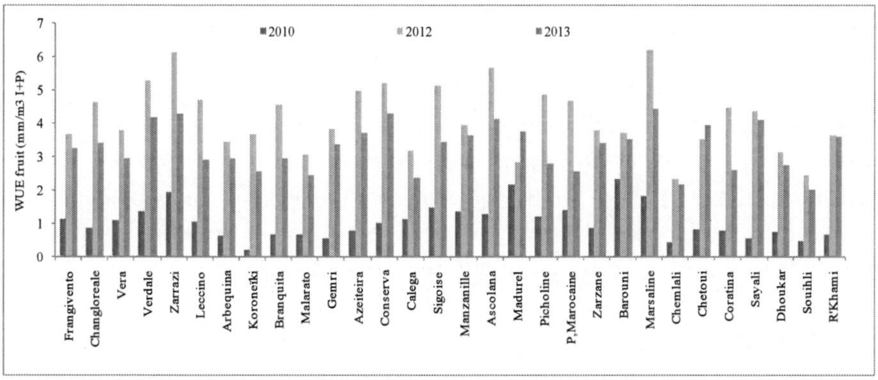

Figure 8. Values of WUE (mm of fruit growth / m^3 I+P) obtained during the compaigns 2010, 2012 and 2013 following the variety.

WUE_{fr} changed also following the stage of development (Table 3). High efficencies were recorded during the early period of growth (stage 1) reaching averages of 12.1 in 2010, 9.6 in 2012 and 6.9 mm/m^3 I+P in 2013, for seasonal supplies of 5.7 m^3/tree in 2010, 3.41 m^3/tree in 2012 and 3.39 m^3/tree in 2013. Maximum WUE_{fr} were 17.47 mm/m^3 I+P in 2010, 15.08 in 2012 and 11.97 in 2013, recorded for cvs., Branquita, Ascolana and Marsaline. Minimum WUE_{fr} were of 8.23 mm/m^3 I+P in 2010, 4.81 mm/m^3 I+P in 2012 and 3 mm/m^3 I+P mm/m^3 in 2013 recorded for oil olive cvs., Chemlali, Koroneiki and Souihli. Efficiencies determined during stage 2 were lower than those observed during the previous period, ranging between

13.51 mm/m^3 I+P (Picholine Marocaine) and -7.66 mm/m^3 I+P (Sayali) in 2010, 9.06 mm/m^3 I+P (Zarrazi) and 0 (Franjivento) in 2012 and 7.3 mm/m^3 I+P (Branquita) and 1.03 mm/m^3 I+P in 2013 (Koroneiki).

Table 3. Values of WUE_{fr} (mm of fruit growth / m^3 I+P) obtained during the compaigns 2010, 2012 and 2013 following the stage of olive development

	Stage 1	Stage 2	Stage 3	Seasonal
2010				
Average	12.1	4.7	0.9	1.0
Ecartype	2.7	3.9	0.6	0.5
Max	17.47	13.51	2.20	2.34
Min	8.23	-7.66	0.15	0.20
2012				
Average	9.6	3.9	2.2	4.2
Ecartype	2.8	1.9	0.7	1.0
Max	15.08	9.06	3.86	6.22
Min	4.81	0.00	0.73	2.33
2013				
Average	6.9	3.4	2.0	3.3
Ecartype	1.8	1.6	0.7	0.7
Max	11.97	7.30	3.45	4.45
Min	3.00	1.03	0.06	1.99

Average values were of 4.7 in 2010, 3.9 in 2012 and 3.4 mm/m^3 I+P in 2013. The lowest WUE_{fr} were recorded during stage 3 with averages of 0.9 mm/m^3 I+P in 2010, 2.2 in 2012 and 2.0 in 2013. Maximums of 2.20, 3.86 and 3.45 mm/m^3 I+P recorded in 2010, 2012 and 2013, respectively were observed for cvs., Barouni, Koroneiki and Sayali. WUE_{fr} relative to cultivars and following the fruit growing stages are reported in figure 9.

Classification of varieties on the basis of their average WUE_{fr} is presented in Table 4. Most table olive fruits are classified in group 2 and oil olive varieties in group 3, but this observation can't be generalized for all table/oil olive varieties.

3.3.4. Olive Production and WUEp

Olive production varied consistently among varieties and between years due to alternate bearing. In 2010, only 20% of trees produced fruits with average productions ranging between 4.1 kg/tree and 12 kg/tree, resulting in low water use efficiencies, varying between 0.7 and 2.1 kg/m^3.

In 2012, 67% of trees produced fruits and productions were variable depending on cultivar. Maximum and minimum productions of 34.3 kg/tree and 10 kg/tree were recorded for cvs., Chemlali and Koroneiki, respectively.

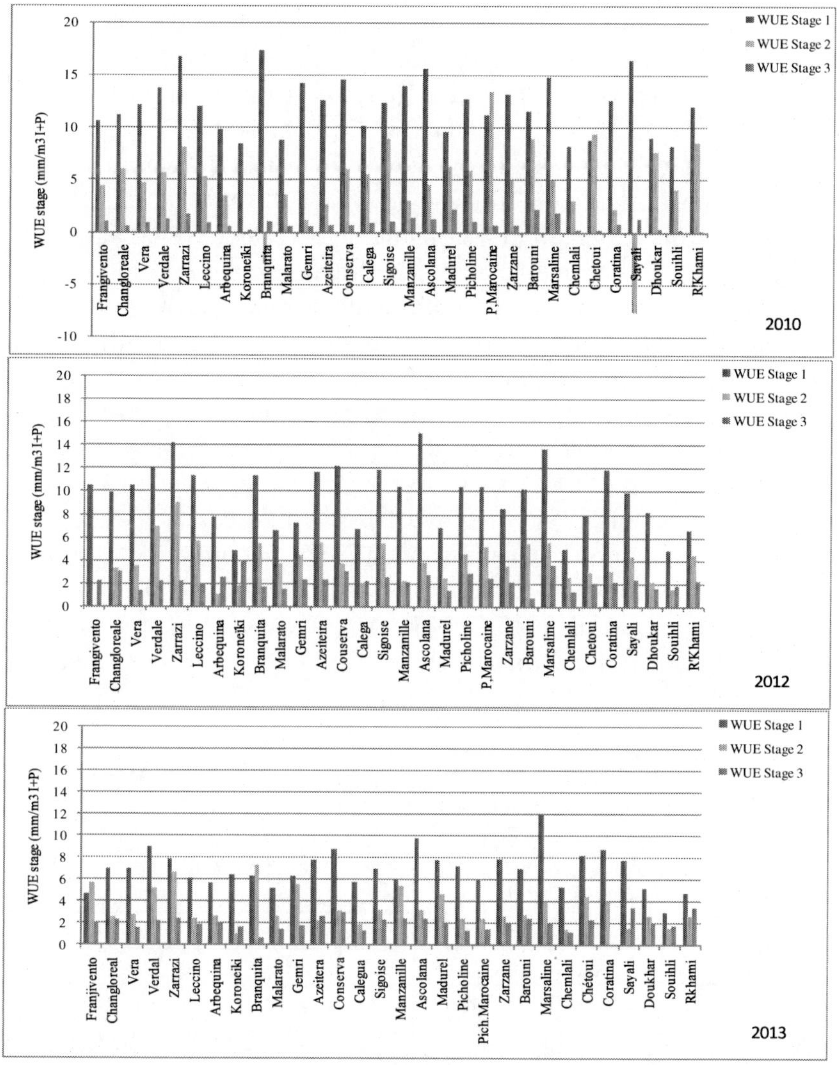

Figure 9. Water use efficiency (WUE_{fr}, mm of fruit growth/m^3) following the year and the fruit growing stages.

Table 4. Classification of varieties on the basis of their WUE_{fr} (mm/m^3)

Class 1	Class 2	Class 3
WUE_{fr} of up of 2.0 mm/m^3	WUE_{fr} ranging from 1.0 to 2.0 mm/m^3	WUE_{fr} < 1 mm/m^3
Barouni, Madurel	Franjivento, Vera, Verdal, Zarrazi, Conserva, Calega, Sigoise, Ascolana, Manzanille, Picholine, Picholine Marocaine, Marsaline.	Rkhami, Souihli, Doukhar, Sayali, Coratina, Chemlali, Zarzane, Azeitera, Gemri, Malarato, Branquita, Koroneiki, Arbequina, Changloreal

In 2013, 73% of trees produced fruits with averages ranging between 2.0 kg (Sayali) and 26.25 kg per tree (Calegua). Values are presented in Table 5. Average production recorded for the three growing seasons was 4.8 kg/tree. The lowest values were observed for the three years of monitoring for cvs., Barouni (0-1.17-0 kg), Doukhar (0-2.7-7.25 kg/tree), Branquita (0-1.3-4.5 kg), Changloreal (0-1-0 kg/tree), Sigoise (0-0.3-11.5 kg/tree), Conserva (0-1.2-0 kg/tree), Dahbia (0-0.3-13 kg/tree), Manzanille (0-0.1-0 kg/tree), and Rkhami (0-1.4-11 kg/tree). Average yield increased from 473 kg/ha in 2010 to 1275 kg/ha in 2012 and 2269 kg/ha in 2013, providing an average value of 1338 kg/ha for the overall experimental period (for the three growing seasons). Largest differences were recorded for cvs., Chemlali, Malarato and Calega. Maximum yield was that of cv., Calegua (4638 kg/ha) and the minimum one for Manzanille (9.3 kg/ha).

Water use efficiences for olive production determined for each year of monitoring and for the overall experimental period are presented in Table 5. Results showed an increase of the average WUE_p from 1.49 kg/m^3 for 2010, to 2.02 kg/m^3 for 2012 and to 3.28 kg/m^3 for 2013, for seasonal water supplies of 5.7 m^3/tree in 2010, 3.44 m^3/tree in 2012 and 3.39 m^3/tree in 2013. Averages were determined on the basis of varieties that produced fruits on each year of monitoring. Highest WUE_p values were recorded for cvs., Chemlali in 2010 (2.11 kg/m^3) and 2012 (10.06 kg/m^3) and Calega in 2013 (7.74 kg/m^3), while the lowest WUE_p were observed for cvs., Arbequina in 2010 (0.72 kg/m^3), Manzanille in 2012 (0.03 kg/m^3) and Sayali in 2013 (0.59 kg/m^3).

Table 5. Productions (kg/tree) and water use efficiencies (WUE_p, kg/m^3) following variety

	P 2010	WUE 2010	P 2012	WUE 2012	P 2013	WUE 2013	Average Prod.	WUEp	Ecartype
Barouni	0		1.17	0.34	0		0.39	0.11	0.68
Calega	8.8	1.54	15	4.40	26.25	7.74	16.68	4.56	8.85
Coratina	0		7.3	2.14	0		2.43	0.71	4.21
Malarato	8.2	1.44	4.3	1.26	25	7.37	12.50	3.36	11.00
Dhoukar	0		2.7	0.79	7.25	2.14	3.32	0.98	3.66
Souihli	0		8.3	2.43	6	1.77	4.77	1.40	4.29
Chemlali	12	2.11	34.3	10.06	2.67	0.79	16.32	4.32	16.25
Branquita	0		1.3	0.38	4.5	1.33	1.93	0.57	2.32
Picholine	0		7.3	2.14	12	3.54	6.43	1.89	6.05
Chétoui	7.9	1.39	9.3	2.73	0		5.73	1.37	5.01
Changloreal	0		1	0.29	0		0.33	0.10	0.58
Arbequina	4.1	0.72	9	2.64	3.5	1.03	5.53	1.46	3.02
Sigoise	0		0.3	0.09	11.5	3.39	3.93	1.16	6.55
Conserva	0		1.2	0.35	0		0.40	0.12	0.69
Dahbia	0		0.3	0.09	13	3.83	4.43	1.31	7.42
Manzanille	0		0.1	0.03	0		0.03	0.01	0.06
Madurel	0		11.7	3.43	0		3.90	1.14	6.75
Rkhami	0		1.4	0.41	11	3.24	4.13	1.22	5.99
Souihli	0		8.3		6		4.77		4.29
Koroneiki	10	1.75	13.3	3.90	0		7.77	1.88	6.93

Table 5. (Continued)

	P 2010	WUE 2010	P 2012	WUE 2012	P 2013	WUE 2013	Average Prod.	WUEp	Ecartype
P. Marocaine	0		0		13	3.83	4.33	1.28	7.51
Zarzane	0		0		13.66	4.03	4.55	1.34	7.89
Leccino	0		0		19	5.60	6.33	1.87	10.97
Azeitera	0		0		8	2.36	2.67	0.79	4.62
Marsaline	0		0		12	3.54	4.00	1.18	6.93
Sayali	0		0		2	0.59	0.67	0.20	1.15
Beldi	0		0		11	3.24	3.67	1.08	6.35
Vera	0		0		7	2.06	2.33	0.69	4.04
Verdal	0		0		12	3.54	4.00	1.18	6.93
Chemchali	0		0		18.5	5.46	6.17	1.82	10.68
Average Values	**1.7**	**1.49**	**4.6**	**2.02**	**8.2**	**3.28**	**4.8**	**1.35**	**5.7**
Ecartype	**3.6**	**0.46**	**7.3**	**2.34**	**7.5**	**1.94**	**4.1**	**1.08**	**3.6**
Max WUE		**2.11**		**10.06**		**7.74**		**4.56**	
Min WUE		**0.72**		**0.03**		**0.59**		**0.01**	
Trees with fruits	**6**		**20**		**22**				
%	**20**		**67**		**73**				

Table 6. Classification of varieties on the basis of their WUE_p (kg/m^3) for olive production

Class 1	Class 2	Class 3
WUE of up of 3.0 kg/m^3	WUE ranging from 1.0 to 2.0 kg/m^3	WUE < 1 kg/m^3
Calegua, Chemlali, Malarato	Picholine, Souihli, Chemchali, Verdal, Beldi, Marsaline, Leccino, Zarzane, Picholine Marocaine, Koroneiki, Rkhami, Madurel, Dahbia, Sigoise, Arbequina, Chétoui.	Barouni, Coratina, Doukhar, Branquita, Changloreal, Conserva, Manzanille, Azeitera, Sayali, Vera.

For the overall plantation, average WUE_p was 2.35 kg/m^3 with maximum value of 4.56 kg/m^3 (Calegua) and minimum of 0.01 kg/m^3 (Manzanille).

Classification of varieties on the basis of their average WUE_p is presented in Table 6. Cultivars Verdal, Picholine Marocaine and Marsaline are classified in group 2 as they provided values of WUE for both fruit growth (WUE_{fr}, mm/m^3) and olive production (WUE_p, kg/m^3) ranging between 1 and 2. The cv., Branquita is classified in group 3, it provided WUE < 1 for both fruit growth and olive production.

3.3.5. Water Use Efficiencies Interrelationships

Values of WUE_{fr} obtained for the different stages of fruit development are well correlated to the increases of fruit size exceeding 30% during stages 1 and 2, with r coefficients of 0.93 and 0.73, respectively (Figure 10).

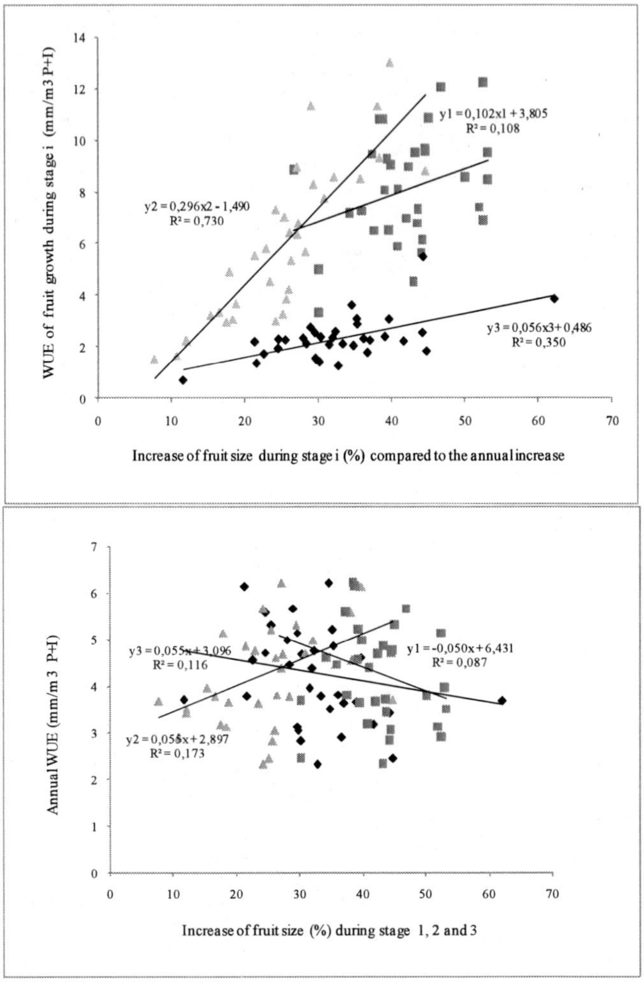

Figure 10. Relationships between fruit diameter increases (%) and WUE_{fr} (mm/m^3 I+P) depending on stage of fruit development.

But when high values of WUE_{fr} are obtained during the different growing periods (2 and 3), the annual WUE_{fr} is not necessary important. Correlations established between annual fruit increases and WUE_{fr} provided correlative coefficients ranging between 0.05 and 0.33 depending on the stage of development.

WUE_{fr} is negatively correlated to soil coverage. More vigorous is the tree (> 15-18% of soil coverage), less efficient is the use of water for fruit growth. Results showed also that

WUE_{fr} is negatively correlated to olive production. This means that when the available water is used for fruit growth, low production level per tree is expected (Figure 11).

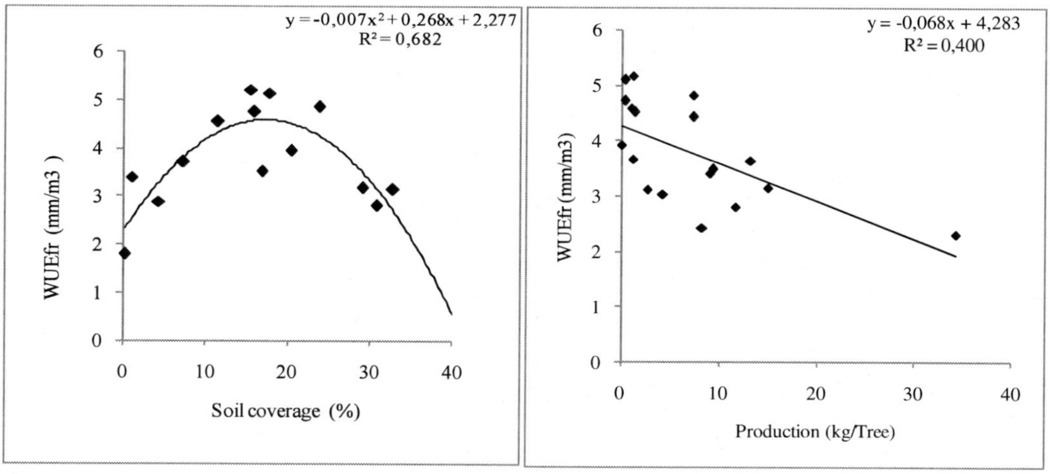

Figure 11. Relationships between WUE_{fr} (mm/m^3 I+P), soil coverage (%) and olive production (kg/tree).

When correlating soil coverage to olive production, different responses were obtained following the year as shown in figure 12. In 2010, a polynomial function was observed with optimum soil coverage around 30% and production of about 10-12 kg/tree. In 2012, linear positive correlations were found with R^2 of 0.83 when production exceeded 8 kg/tree and 0.51 for trees yielding less than 8 kg/tree. Production is also correlated, linearly but negatively, to the annual increase of fruit size with R^2 of 0.40.

Water use efficiency for olive production is also correlated to soil coverage, L/D ratio and fruit diameter as shown in figure 13, with optimum soil coverage of 30% and R^2 values of 0.76 when correlating L/D to WUE_p and 0.19 for the correlation observed between fruit diameter and WUE_p.

Correlations established between growth parameters and production showed a great variability which is depending on fruit load and alternate bearing year. Particularly, the correlation found between WUE_{fr} and production, showed that the efficacity of using water for fruit growth decreased when the production level and soil coverage increased, stipulating that more vigorous is the tree, less efficient is the use of water for fruit enlargement. Optimum soil coverage ranges between 15% and 18% for 10 years old orchard. More investigation is needed to well correlate shoot elongation, soil coverage, production of fruits and soil water content. On the other hand, classification of varieties on the basis of WUE for olive production and fruit size showed maximum values for cvs., Calegua, Chemlali, Malarato, of up to 3 kg/m^3 P+I, while minimum values were recorded for cvs., Manzanille, Coratina, Azeitera and Conserva. The cvs., Verdal, Picholine Marocaine and Marsaline provided average efficencies varying between 2 and 3 for both olive production (kg/m^3P+I) and fruit growth (mm/m^3 P+I).

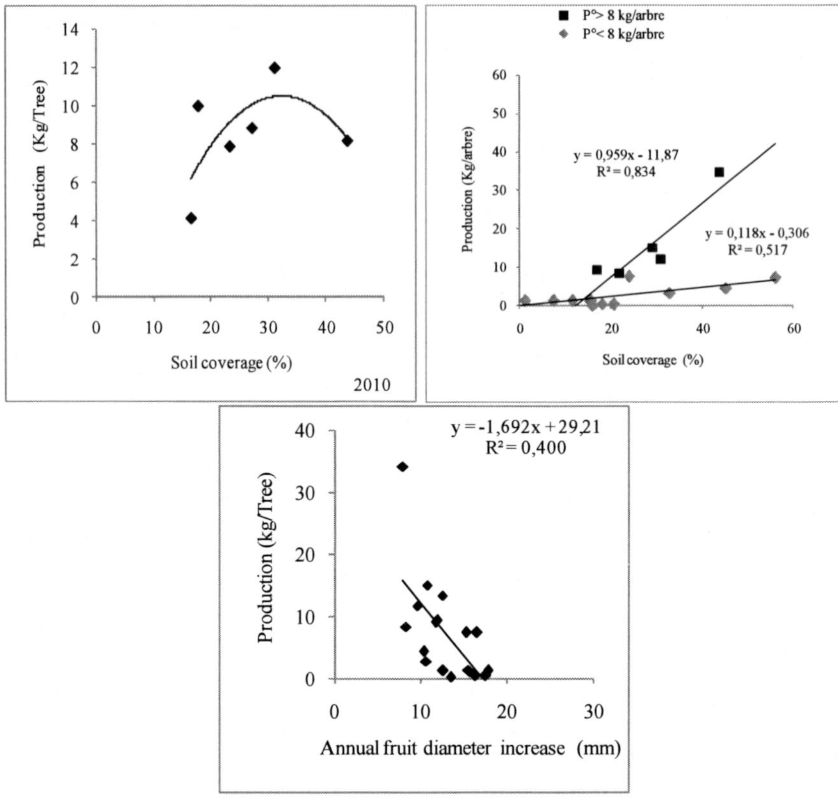

Figure 12. Relationships established between soil coverage (%) and fruit increases and olive production (kg/tree).

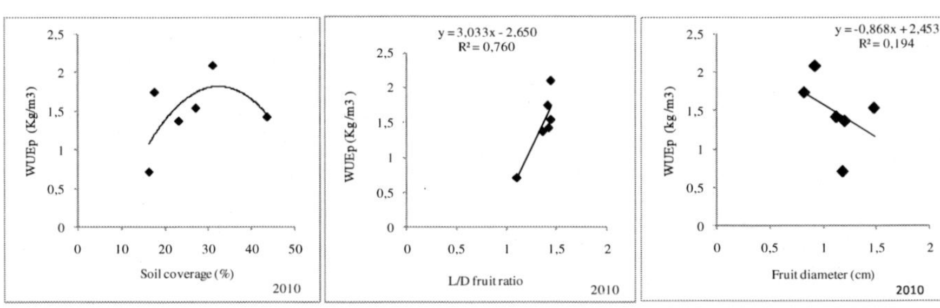

Figure 13. Relationships between water use efficiency for olive production (WUE_p, kg /m^3 I+P), soil coverage and fruit diameter.

4. DISCUSSION

This research work was carried out concomitantly with the investigations made all over the Mediterranean basin on irrigation management of olive orchards. Indeed, as interest increased nowadays for olive and because most olive groves are cultivated under water shortage conditions, particularly in the southern side of Mediterranea, it becomes necessary for us to study judiciously the agronomical and physiological responses of local and foreign cultivars to water application and intensive cultivation system: (1) under the same environmental conditions for all cultivars and (2) why not –in the future- in different growing areas, but simultanuously. In this experiment we studied the behaviour of panoply of varieties cultivated in northern Tunisia, in the same orchard and under drip irrigation, *i.e.,* case (1). Results showed that the doses of water applied (5.7 m^3/tree in 2010, 3.41 m^3/tree in 2012 and 3.39 m^3/tree in 2013) from fruit set to the stage of olive maturation, meeting more than 50% of the crop water needs (71% ET_c in 2010 and 56% in 2012 and 2013), allowed humectation of the tranches of soil (0-60cm) where roots are growing. However, when examining the soil profil, large differences in water contents were observed following the site of measurement, depth and the stage of fruit development. For most growing stages, the amounts of irrigation supplies were correctly applied; but it was not the case during the period that follows flowering. Indeed, low values of water contents were recorded, leading to a substantial water deficit by end of May and in June, *i.e.,* during the period of rapid fruit growth and embryo development. At the same time, some olive varieties showed a shrinkage of their fruit diameters (-3%to -21%). Therefore, it appears that an increase of the inflows is required during this period, which can be considered –in our area- as a 'critical stage'. These results are concordant with those published by Masmoudi-Charfi (2013a) about the distribution of water around the olive trees and those of Rallo (1998), precising their sensitive stages when olive trees are growing in the northern area of Mediterranea. The shema presented by this author showes a delay of about one month compared to our experiment. At this stage of our study, more investigation is needed to specify the amount of water consumed really by these varieties during each stage of fruit development, for example, by measuring the sap flow. Such work was done on young olive trees cultivated in northern Tunisia by Masmoudi-Charfi et al., (2012), providing water consumption ranging between 0.09%ET_0 and 20%ET_0 following the season, with maximum sap flux value recorded in August. The study of Abid-Karray (2006) was made on older trees growing under the arid climate of central Tunisia, showing higher amounts of sap flux, of up to 40 l/day. In the proposed further study, particular attention would be attributed to the distribution of roots and their anatomy (Charfi, 1990; Masmoudi-Charfi and Ben Mechlia, 2009) which affect the responses of trees as mentioned by Fernandez et al., (2003).

The panoply of varieties experimented in Nabeul behaved differently face to the intensive growing conditions and water application, although some of them were classified in the same category and have similar vigor. Particularly, the soil coverage showed a great variability among varieties and from year to another. High percentages of up to 30% were observed for cvs., Coratina (50.1%), Chemlali (39.5%), Zarzane 37.5%), Madurel (32.7%), Malarato (46.6%). Low values of less than 15% were recorded for cvs., Franjivento (6.3%) - Changloreal (9.7%) – Branquita (11.8%) - Souri Liban (8.4%) – Gemri (7.7%) - Azeitera (9.3%) - Besbassi (12.4%). A consistant increase of soil coverage was observed between 2010 and 2013 for most varieties with maximums recorded for Doukhar (x2.3), Franjivento (x2.2) and Picholine Marocaine (x1.9). Average value increased from 16.7% in 2010, to 21.8% in 2012 and to 24.2% in 2013. This is an impotant result because the increase of soil coverage is normally accompanied by an increase of the leaf area and thus an increase of the amount of assimilates available to fruits (Proeitti and Tombesi, 1996; Fernandez and Moreno, 1999; Connor and Fereres, 2005). Concomitent measurements of leaf area and estimation of the amount of starch, can lead to more precision about the relationship between the increase of soil coverage and the availability of assimilates to fruits.

A great variability was recorded between varieties in fruit dynamic and in annual gains. Highest ratios observed between fruit size at harvest and that recorded at fruit set concerned the cvs., *Madurel* in 2010 (x3), *Barouni* in 2010 (x2.5), *Chemlali* in 2012 (x3) and *Doukhar* in 2012 (x2.5). The lowest ratios (x1) were those recorded for cvs., *Koroneiki* (2010) and *Sayali* (2012). The largest differences between years were recorded for cvs., *Branquita, Gemri and Sayali,* which were apparently more sensitive to water application and alternate bearing than the cvs., *Changloreal, Leccino, Calegua, Chemlali and Souihli,* showing slight dispearences. In addition to fruit size at fruit set and at harvest, the regular monitoring of fruit diameter from fruit set to the ripening stage, showed that fruits grow continuously, but following different trends and models depending on variety. Some of them like the cvs., *Changloreal, Koroneiki, Azeitera, Conserva, Manzanille, Picholine Marocaine and Zarzane* ensured a typical growth pattern, following three growing periods as observed by Masmoudi-Charfi (2013) for Picholine, Manzanille, Meski and Chétoui cvs., cultivated in north of Tunisia under drip irrigation. The cvs., *Leccino, Gemri, Sigoise, Madurel, Barouni, Marsaline, Chétoui, Sayali, Verdal, Calegua,* showed 'atypical' model with a unique rapid growth period. This result corroborates that of Boulouha (1986) reporting one period of growth for irrigated olive trees of cv., *Picholine Marocaine* from early spring to late summer. However, in spite of these differences, maximum fruit gains were always recorded during the early stage of fruit development, *i.e.,* just after fruit set. Results showed also strong plasticity of some cultivars although they were cultivated under some water shortage conditions particularly the cv., *Picholine-* adjusting its responses to the available amounts of water and to the alternate bearing conditions. These results corroborate also Wahbi et al., (2005) and Mezghani Ayachi et al., (2012) whom reported significant differences between cvs., face to water use following their characteristics and the year's conditions. Revealing such differences are very important, showing that olive varieties didn't profit from water supplies similarly. Some of them valorise water during their early fruit growth process, just after fruit set, while others valorize water more efficiently after pit hardening. On the basis of these results, we can conclude that varieties should be judiciously associated when planting multi-varietal orchards to get harmonious groves, correct fruit development and to allow good estimation of the irrigation water needs. Particular attention should be attributed to the 'sensitive' cvs., like *Dahbia, Lucques and Koroneiki* showing some shrinkage of their fruit size when water becomes less available. Other cvs., stopped growing for about 20-days long period from end of June to 10 of July due to water shortage. Previous researches (Rallo and Suarez, 1989; Proietti and Tombesi, 1996) showed similar results and reported a significant reduction of fruit size and their number on trees cultivated under limited water conditions, due to low translocation of dry matter. According to Lavee et al., (1990), water controls the transfert of carbohydrates to the different organs of the tree, ensuring shoot elongation, enlargement of fruits and controlling the processes of flowering, fruit set, fruit drop, and alternate bearing.

Enlargement and shrinkage of fruits are also highly dependent on the auxin/gibberellin ratio (Chehab, 2009), fruit load and the proximity of fruits from the sites of assimilate's production (mature leaves). This was largely discussed in Suarez et al., (1984), Mickelakis, (1995); Proietti and Tombesi (1996), Lavee, (1997); Fernandez and Moreno (1999); Michelakis (2000), Patumi et al., (2002); Castillo-Llanque et al., (2005 and 2008) Tognetti, (2006) and Iniesta et al., (2009), Martin-Vertedor et al., (2011a and 2011b) and Aïachi et al., (2012a). These authors attribute differences between varieties to the well-known competition for water and nutrients between the growing fruits, shoots and roots. This means that after setting, the drupe becomes the major sink and attracts the nearby available substances as well as those located in other sites, increasing strongly the competition for nutrients between the newly formed shoots and the growing fruits, which compete also with roots and become the first responsible on assimilates' partitioning. This is why high yield year is generally followed by a low yield year even under optimal conditions of cultivation. Suarez et al., (1984), Rallo and Suarez (1989), Costes et al., (2000), Mickelakis (1990 and 2000) and D'Andria et al., (1998) supported the hypothesis that crop load of trees interacts with primary growth rhythmicity and more precisely with growth resumption. However, this behavior could not be generalized for all olive varieties because the alternate bearing process is controlled genetically. Some cultivars may present a strict alternance like the cv., Chemlali of Sfax, while others like the cv., Picholine yield correctly every year, in Tunisia. These varieties may exhibit a different behavior when they are cultivated elsewhere, providing higher or lower yields according to their production potential and the growing conditions.

High average productions were recorded for cvs., *Calegua, Malarato, Chemlali and Koroneiki* which are both oil varieties. The amounts of water supplied (71% ET_c in 2010 and 56% in 2012 and 2013) during the fruit growth period were apparently suffisant to ensure such yields, and thus they can be adopted as optimums for these cvs., These results corroborates Grattan et al., (2006) reporting high yields for water amounts of 71-89% $ET_{c,}$ Also, Patumi et al., (2002) noted that a restitution of 66% ET_c was sufficient to achieve good yields of most varieties but this water should be applied judiciously and at correct time. Chalmers et al., (1981) and Fernandez and Moreno (1999) reported that the maintenance of a slight plant water deficit can control more efficiently the excessive vegetative growth and improve the partitioning of carbohydrates to reproductive structures such as fruits. In opposite, cvs., like *Manzanille and Conserva* provided low yields. This result was unexpected because these varieties are vigorous and have an importance soil coverage. For these varieties, we assume that the amount of water was not adequate. Iniesta et al., (2009) and Palease et al., (2010) indicated that excessive water amounts have negative effect on growth and olive production. Experiments carried out in southern Tunisia showed for cv., Chemlali a decrease of the olive production when the annual water supplies exceeded 400 mm (Masmoudi-Charfi et al., 2010).

Water use efficiencies (WUE) for both fruit growth and olive production varied consistently between years and among varieties as a result of the different observed rates of fruit growth and production levels. Highest WUE_p values were recorded for cvs., Chemlali in 2010 (2.11 kg/m^3) and 2012 (10.06 kg/m^3) and Calega in 2013 (7.74 kg/m^3), while the lowest, were observed for cvs., Arbequina in 2010 (0.72 kg/m^3), Manzanille in 2012 (0.03 kg/m^3) and Sayali in 2013 (0.59 kg/m^3). Average WUEp was equal to 2.35 kg/m^3 with maximum of 4.56 kg/m^3 (Calegua) and minimum of 0.01kg/m^3 (Manzanille). For fruit growth, WUE_{fr} increased from 1 mm/m^3 in 2010 to 4.2 mm/m^3 in 2012, but decreased in 2013 reaching 3.3 mm/m^3. Compared to values obtained actually in the irrigated adult plantations of north of Tunisia, these WUE are consided too high and exceed the normes given by the COI (Lavee, 1997). Values observed herein approximate those obtained in 2009 and 2010 in south Tunisia (P=200 mm) for cvs., Manzanille Chétoui, Picholine and Coratina cultivated under drip irrigation, at 7mx7m spacing and receiving an irrigation amount of 20%ET_c (120 mm). For these cvs., WUE_p ranged between 1.6 and 3.9 kg/m^3. For the cv., Chemlali de Sfax, the WUE_p was equal to 1.2 kg/m^3, provided at 50% ET_c (250 mm) (Masmoudi-Charfi et al., 2010; Aïachi-Mezghani et al., 2012). The lowest WUE_p were obtained at 100%ET_c.

So and as reported previously, the response of olive varieties and their ability to valorize water for fruit growth or olive production was found to be dependent on multitude of factors (Proietti and Tombesi, 1996; Michelakis, 2000; Fernandez et al., 2003; Connor and Fereres, 2005; Grattan et al., 2006; Ben Ahmed et al., 2007, Chehab, 2009; Ayachi Mezghani et al., 2012 a and b, Masmoudi-Charfi et al., 2012a...), amongst the importance of the transpiration area and their genetic potential to produce or not high yields.... making the reponse of the tree variable and sometimes unexpected. Sibbett (2002) discussed a range of possible factors to explain the response of varieties, including time of water application and that of harvest. It could be also interesting to speculate on the possible implication of hormonal status, *i.e.,* the auxin/gibberellin ratio and its seasonal changes on the overall olive growth cycle (Proietti and Tombesi, 1996; Chehab et al., 2009). Besides, the response to water is highly dependent on variety adaptation and its aptitude to overcome the lack of water in a given area at a specific time. The cultivar Chemlali, for example, cultivated in Centre and South of Tunisia showed since centuries a great adaptation to local conditions; it vegetates and yields even with low water supplies by developing deep roots and by decreasing its leaf water potential to very low values (-4.5 KPas). Tognetti et al., (2006) observed that the cv., Leccino yielded correctly in Italy for an amount of 33% ET_c, however the cv., Frontoio required higher amount of water to reach a significant yield enhancement. Other factors, non-physiological, like the horticultural practices, the localization of water application, the homogeneity of water distribution... (Lebourdelles, 1977; Cruz-Cond Suarez de Tangil and Fuentes-Cabanas, 1989; Lavee, 1997; Fernandez and Moreno, 1999; Pastor et al., 1998; Ghrab et al., 2003; Pastor 2005; Masmoudi-Charfi et al., 2012b) should be considered. This together, with the previously cited reasons may explain the unexpected results obtained for some varieties.

CONCLUSION

Considering the different targets aimed in the orchards of olive varieties, it can be inferred that continuous water availability from early spring to end autumn period is primordial for successful flowering, adequate new vegetation and high yield in intensive orchards. However, the amounts of water distributed should be well-estimated depending on variety (table and oil varieties) and considering the first priority. For most varieties, adequate yields can be achieved in the northern area with a water contribution of 50-70% ET_c (200-300 mm). Higher irrigation amounts could be advisable during 'On' years. However, the most important factor is the choice of varieties particularly in the multi-varietal orchards. Indeed, there is a need for a more accurate scheduling of water application in such plantations because of the different combinations (region, cultivar) and particularly for the semi-arid and arid regions of the country. To choose varieties, WUE can be considered as a usefull tool to well associate cvs., in this kind of groves depending on the objective. For table olive plantations, WUE_{fr} is well adapated, while WUE_p is suitable for olive oil production orchards. On the other hand, relationships established between all growing variables: soil coverage, olive production, fruit diameter and water use efficiencies showed a particular importance of soil coverage wich varied substantially following cultivars and should be taken into account when determining the crop water needs of multi-varietal olive orchards for a better estimation of irrigation supplies.

ACKNOWLEDGMENTS

Authors wish to thank Dr., Msallem Monji for authorizing experiments in the orchard of Nabeul, Mme Yakoubi Samira and Mrs., Ben Dhiab Ali and Hamadi Ali for their assistance during the experiment. Thanks to Bilel Alaoui for collaboration.

REFERENCES

Abid-Karray J., 2006. Bilan hydrique d'un système de cultures intercalaires (olivier-culture maraîchère) en Tunisie centrale: Approche expérimentale et essai de Modélisation. Thèse de Doctorat (*PhD*). Université de Montpellier II. p. 172.

Aïachi Mezghani, M., Sahli, A., Jebari, A., 2007. Analysis and modelling of the vegetative growth of the olive tree (Olea europaea L.) in rainfall conditions. *Fruits,* 61: 45-56.

Aïachi Mezghani M., A. Sahli, N. Grati, F. Ben Amar, S. Ben Ali, R. Ben Amor, F. Labidi and A. Ouled Amor, 2009. Olive trees characteristics (Olea europaea L.) under intensive condition cultivated in Tunisia (Sfax site). In For a renovate, profitable and competitive Mediterranean olive growing sector. Olivebioteq 2009, Sfax, Tunisie, 15-19 Décembre 2009, Book of proceedings, 443-447.

Aïachi Mezghani M., Masmoudi Charfi, C., Gouiaa, M., Labidi, F., 2012a. Vegetative and reproductive Behaviour of some olive tree cultivars (Olea europaea L.) under deficit irrigation regimes in semi-arid conditions of Central Tunisia. *Scientia Horticulturae*, 146, 143-152.

Aïachi Mezghani M., A. Sahli, F. Labidi, M. Khairi and A. Ouled Amor, 2012b. Study of the Behaviour of Olive Varieties cultivated under Different Tree-trainings: Vegetative and Productive Characteristics. *Acta. Horticulturae,* 949: 439-446.

Allen, R., Pereira, L. S., Raes, D., Smith, M., 1998. Crop evapotranspiration. Guidelines for computing crop water requirements. Irrigation and drainage paper No 56, FAO, Rome.

Androulakis, I. I., 1997. Growth and Development of the Olive (Olea Europaea L.) in Relation to Irrigation, Mineral Nutrition and Fruit Load. The Yearly Vegetative Growth Cycle and the Alternate Bearing. Proceedings on the International seminar on olive growing, 18-24 May, Chania Greece: 27-32.

Bamouh A., Ezzahouani A., ELhatmi S., 2012. Les besoins en eau de l'Olivier sur la base de son taux de couverture du sol dans la zone de Beni Mellal. HTE N° 151 - Mars 2012.

Bchir A., Boussadia O., Lemeur R., Braham M., 2003. Water use in olive orchards estimated by physiologic and climatic methods in Tunisia. *European Scientific Journal*, August, 2013 edition vol.9, No.24 ISSN: 1857 – 7881 (Print) e - ISSN 1857- 7431.

Ben Ahmed, C., Ben Rouina, B., Boukhris, M. M., 2007. Effects of water deficit on olive trees cv. Chemlali under field conditions in arid region in Tunisia. *Sci. Hortic.*, 113: 267-277.

Bonji, G., Palliotti, A., 1994. Olive in Hand book of environmental physiology of fruit crop. Eds. B. Schaffe and P.C., Anderson CRC Press Inc. 165-187.

Boulouha, B., 1995. Contribution à l'amélioration de la productivité et de la régularité de la production chez l'olivier (Olea europaea) Picholine Marocaine'. *Olivae*, 58: 54-57.

Castillo-Llanque, F. J., Rapoport, H. F., Navarro, C., 2005. Caracterizacion de la ramification del olivo: influencia de differentes fechas de recolleccion y estados de carga. In: Congreso Iberico de Ciencias Horticolas, May 2005, Porto, Portugal, 1-7.

Castillo-Llanque, F. J., Rapoport, H. F, Navarro, C., 2008. Interaction between shoot growth and reproductive behavior in olive trees. *Acta. Hortic.*, 791: 453-457.

Chalmers, D. J., Mitchell, P. D., Vanheek, L., 1981. Control of peach-tree growth and productivity by regulated water-supply, tree density and summer pruning. *J. Am. Soc. Hortic. Sci.,* 106 : 307-312.

Charfi, C., 1990. Multiplication de l'olivier par bouturage herbacé dans les caissons chauds. Mémoire de fin d'études du cycle d'ingénieur. INAT. 60 pages.

Chehab, H., Mechri, B., Ben Mariem, F., Hammami, M., Ben ElHadj, S., Braham, M., 2009. Effect of different irrigation regimes on carbohydrate partitioning in leaves and wood of two table olive cultivars (Olea europaea L., cv. Meski and Picholine). *Agric. Water Manage.*, 96: 293-298.

Connor, D. J., Fereres, E., 2005. The Physiology of Adaptation and yield Expression in Olive. *Horticultural Review*, 31, 155-229.

Costes, E., Fournier, D., D., Salles, J. C., 2000. Changes in primary and secondary growth as influence by crop load in'Fantasme" apricot trees. *J. Hortic. Sci. Biotechnol.*, 75: 510-519.

Cruz-Conde Suarez de Tangil, J., Fuentes-Cabanas M., 1989. Irrigation goutte à goutte de l'oliveraie: doses d'eau. Olivae n 25, pp 24-25.

D'Andria, R., Tognetti, R., Morales-Sillero, A., Fernandez, J. E., Sebastiani, L., Troncoso, A., 2008. Deficit irrigation and fertirrigation practices in olive growing: Convergences and divergences in two case studies. *Plant. Biosyst.,* 142:138-148.

DGPA, 2006. Direction générale de la production agricole. Indicateurs macro-économiques et agricole. E 108. Doc 4.

Fernandez, J. E., Palomo, M., Diaz-Espejo, A., Giron, I., 2003. Influence of partial soil wetting on water relation parameters of the olive tree. *Agronomie,* 23: 545-552.

Fernandez, J. E., 2006. Irrigation Management in Olive. Olivebioteq., 295- 305.

Fernandez, J. E., Diaz-Espejo, A., Infante, J. M., Duran, P., Palomo, M. J., Chamorro, V., Giron, I. F., Villagarcia, L., 2006. Water relations and gas exchange in olive trees under regulated deficit irrigation and partial root zone drying. *Plant and Soil*, 284: 273-291.

Fernandez, J. E., Moreno, F., 1999. Water use by olive tree. *J. Crop. Prod.*, 2, 101-162.

Fernandez, J. E., Palomo, M., Diaz-Espejo, A., Giron, I. F., 2003. Influence of partial soil wetting water relation parameters of the olive tree. *Agronomie,* 23: 545-305.

Fernandez, J. E., Diaz.Espejo, A., Infante, J. M., Duran, P., Palomo, M. J., Chamorro, V., Giron, I. F., Agarcia, L., 2006. Water relation and gaz exchange in olive trees under regulated deficit irrigation and partial rootzone drying. *Plant Soil*, 284: 273-291.

Forshey, C. G., Elfving, D. C., 1989. The relationship between vegetative growth and fruiting in apple trees. *Hortic. Rev.*, 11, 229-87.

Ghrab, M., Gargouri, K., Benthaer, H., 2003. Influence de l'irrigation goutte a goutte par des eaux chargées sur un sol leger. In: Camarda D. (ed.), Grassini L. (ed.). Local resources and global trades: Environments and agriculture in the Mediterranean region. Bari : CIHEAM, 2003. p. 63-67 (Options Méditerranéennes: Série A. Séminaires Méditerranéens; n. 57).

Ghrab M., Gargouri K., Bentaher H., Chartzoulakis K., Ayadi M., Ben Mimoun M., Masmoudi M. M., Ben Mechlia N., Psarras G., 2013. Water relations and yield of olive tree (cv. Chemlali) in response to partial root-zone drying (PRD) irrigation technique and salinity under arid climate. *Agricultural Water Management,* 123: 1– 11.

Goldhamer, D. A., Dunai, J., Ferguson, L. F., 1994. Irrigation requirements of olive trees and responses to sustained deficit irrigation. *Acta. Hortic.*, 356: 551-559.

Goldhamer, D. A.,1999. Regulated deficit irrigation for California canning olives. *Acta. Hortic.*, 474: 369-372.

Grattan, S. R., Connel, J. H., Polito, V. S., 2006. Tree irrigation levels for optimum chemical and sensory properties of olive oil. *Horticultural Science,* 41: 427-432.

Gucci, R., Lodolini, E., Rapoport, H. F., 2007. Productivity of olive trees with different water status and crop load. *J. Hortic. Sci. Biotech.*, 82, 648-656.

Hamza, M., 2009. La politique de l'eau en Tunisie. Conférence régionale sur la gouvernance de l'eau. Echange d'expérience entre l'OCDE et les pays arabes. CITET. Direction Générale des Ressources en Eau. Tunisie, 8-9 Juillet 2009.

Iniesta, F., Testi, L., Orgaz, F., Villalobos, F. J., 2009. The effects of regulated and continuous deficit irrigation on the water use, growth and yield of olive trees. *Eur. J. Agron.*, 25: 258-265.

Larbi, A., Ayadi, M., Ben Dhiab, A., Msallem, M., 2009. Comparative Study of Tunisian and Foreign Olive Cultivars Sustainability for High Density Planting System. Olivebioteq, Sfax-Tunisia, 177-181.

Lavee, S., 1997. Biologie et physiologie de l'olivier. In: Plaza, S., Janes, A. (Eds.). Encyclopédie Mondiale de l'Olivier. Servers Editorials Estudi Balm, Barcelona, Spain, pp. 61-110.

Lavee, S., Rallo, L., Rapoport, H. F., Troncoso, A., 1999. The floral biology of the olive. II. The effect of inflorescence load and distribution per shoot on fruit set and load. *Sci. Hortic.*, 66: 267-277.

Le Bourdelles, J., 1977. Irrigation par goutte à goutte en Oléiculture. Principe de la méthode, Installations et Fonctionnement. Olea, Juin 1977.

Martin-Vertedor, A. I., Perez Rodriguez, J. M., Prieto Losada, M. H., Fereres Castiel, E., 2011a. Interactive responses to water deficits and crop load in olive (Olea Europaea L., cv.Morisca). I - growth and water relations. *Agric. Water Manage.*, 98: 941-949.

Martin-Vertedor, A. I., Perez Rodriguez, J. M., Prieto Losada, M. H., Fereres Castiel, E., 2011b. Interactive responses to water deficits and crop load in olive (Olea Europaea L.,cv. Morisca). II-Water use, fruit and oil yield. *Agric. Water Manage.,* 98, 950-958.

Masmoudi-Charfi, C., Masmoudi, M. M., Ben Mechlia, N., 2004. Irrigation de l'Olivier: Cas des jeunes plantations intensives. *Revue Ezzaitouna*, 10 (1-2), 37-51.

Masmoudi-Charfi, C., 2006. Gestion de l'irrigation dans les plantations d'olivier. Document Technique n°3. Institut de l'Olivier. pp. 13. (Technical paper).

Masmoudi-Charfi, C., Msallem, M., Say, M. B., 2006. Plantations irriguées d'olivier. - Document Technique n°2. Institut de l'Olivier. pp. 7. (Technical paper).

Masmoudi-Charfi, C., Ayachi-Mezghani, M., Gouia, M., Laabidi, F., Ben Reguaya, S., Ouled Amor, A., Bousnina, M., 2010. Water relations of olive trees cultivated under deficit irrigation regimes. Sci.Horti., 125: 573–578.

Masmoudi-Charfi C., Masmoudi M. M., Abid Karray J., Ben Mechlia N., 2012a. The Sap Flow Technique: A Precise Means to Estimate Water Consumption of Young Olive Trees

(Olea Europaea L.). Book Chapter, 2. Irrigation Management Technologies and Environmentam Impact. Ed., Hossain Ali. Nova Sciences Publishers, Hauppage NY. https://www.novapublishers.com/catalog/product_info.php

Masmoudi Charfi C., Gargouri K., Habaieb H., Daghari H., Abid-Karray J., Rhouma A., 2012b. Manuel d'Irrigation de l'Olivier. Techniques et Applications 110 pages. Ed. Institut de l'Olivier.

Masmoudi Charfi C., 2013a. Quantitative analysis of soil water content in young drip irrigated olive orchards. *Advances in Horticultural Sciences,* 26 (3-4).

Masmoudi-Charfi C., 2013b. Growth of Young Olive Trees: Water Requirements in Relation to Canopy and Root Development. Special Issue on Plant Growth and Development. American Journal of Plant Sciences (AJPS, ISSN: 2158-2750), ajps, 2013, Volume 4, 1316-1344. Published Online July 2013. (http://www.scirp.org/journal/ajps).

Masmoudi Charfi C., Habaieb H., 2014. Rainfall Distribution Functions for Irrigation Scheduling: Calculation Procedures Following Site of Olive (Olea europaea L.) Cultivation and Growing Periods. *American Journal of Plant Sciences,* 2014, 5, 2094-2133. Published Online June 2014 in SciRes. http://www.scirp.org/journal/ajps. http://dx.doi.org/10.4236/ ajps.2014.513224

Melgar, J. C., Mohamed, Y., Navarro, C., Parra, M. A., Benlloch, M., Fernandez-Escobar, R., 2008. Long-term growth and yield responses of olive trees to different irrigation regimes. *Agric. Water Manage*., 95, 968-972.

Michelakis, N., 1990. Yield response of table and oil olive varieties to different water use levels under drip irrigation conditions. *Acta. Hortic.,* 286: 271-274.

Michelakis, N., 1995. Effet des disponibilités en eau sur la croissance et le rendement des oliviers. *Olivae*, 56: 29-39.

Michelakis, N., Vougioucalou, E., 1988. Water use, root and top growth of olive trees for different methods of irrigation and levels of soil water potential. *Olea,* 19, 17-31.

Michelakis, N., 2000. Water requirements of olive tree on the various vegetative stages. Proceedings of the International Course on water management and irrigation of olive orchards. Cyprus.

Moriana, A., Orgaz, F., Pastor, M., Fereres, E., 2003. Yield responses of a mature olive orchard to water deficits. *J. Am. Soc. Hortic. Sci.,* 128: 425-431.

Orgaz, F., Testi, L., Villalobos, F. J., Fereres, E., 2006. Water requirements of olive orchards. Determination of crop coefficients for irrigation scheduling. *Irrigation Science*, 24: 77-84.

Palease, A. M., Nuzzo, V., Favati, F., Pietrafesa, A., Celano, G., Xiloyannis, C., 2010. Effects of water deficit on the vegetative response, yield and oil quality of olive trees (Olea europaea L.,cv.. Coratina) grown under intensive cultivation. *Sci. Hortic*., 125: 222-229.

Palomo, M. J., Moreno, F., Fernandez, J. E., Diaz-Espejo, A., Giron, I. F., 2002. Determining water consumption in olive orchards using the water balance approach. *Agricultural water management*, 55:15-35.

Pastor, M., Hidalgo J., Castro J., 1998. Irrigation des cultures oléicoles dans la région de la Loma (province de Jaén) Olivae N) 71- Avril 1998, pp 39-49.

Pastor, M., 2005. Cultivo del olivo con riego localizado. In: Junta de Andalucia y Ediciones Mundi-Prensa (Eds), 783 pp plus CD, ISBN 84-8476-229-7.

Patumi, M., D'Andria, R., Marsilio, V., Fontanazza, G., Morelli, G., Lanza, B., 2002. Olive and olive oil quality after intensive monocone olive growing (Olea europaea L., cv. Kalamata) in different irrigation regime. *Food Chem.,* 77: 27-34.

Pérez-Lopez, D., Ribas, F., Moriana, A., Olmedilla, N., De Juan, A., 2007. The effect of irrigation schedules on the water relations and growth of a young olive (Olea europaea L.) orchard. *Agric. Water Manag*., 89: 297-304.

Proietti, P., Tombesi, A., 1996. Translocation of assimilates and source-sink influences on productive characteristics of the olive tree. *Adv. Hortic. Sci.,* 10: 11-14.

Rallo, L., Suarez, M. P., 1989. Seasonnal distribution of dry matter within the olive fruit - bearing limb. *Adv. Hortic. Sci*., 3:, 55-59.

Roméro, M. P., Tovar, M. J., Motilva, M. J., 2002. Changes in the HPLC phenolic profile of virgin olive oil from young trees (Olea europaea L.,cv. Arbequina) grown under different deficit irrigation strategies. *J. Agri. Food Chem.,* 50: 5349-5354.

Sibbett, S., 2002. Orchard management. *Aust. Olive Grower*, 29:7-9.

Suarez, M. P., Fernandez-Escobar, R., Rallo, L., 1984. Competition among fruits in olive. II. Influence of inflorescence or fruit thinning and cross-pollination on fruit set components and crop efficiency. *Acta. Hortic*., 149, 131-143.

Testi, L., Villalobos, F. J., Orgaz, F., Fereres, E., 2006. Water requirements of olive orchards: I simulation of daily evapotranspiration for scenario analysis. *Irrigation Science*, 24: 69-76.

Tognetti, R., D'Andria, R., Morelli, G., Alvino, A., 2005. The effect of deficit irrigation on seasonal variations of plant water use in Olea europaea L. *Plant Soil*, 273: 139-155.

Tognetti, R., D'Andria, R., Lavini, A., Morelli, G., 2006. The effect of deficit irrigation on crop yield and vegetative development of Olea europaea L. (cvs.Frantoio and Leccino). *Eur. J. Agron*., 25: 356-364.

Trigui A., Msallem M., Yengui A., Belguith H., Khecherem J., Meliène A., Malek S., Bousselmi A., Samet A., Trabelsi E.B., 2002. Oliviers de Tunisie: Catalogue des variétés autochtones et types locaux. Volume I [Identification variétale and caractérisation morpho-pomologique des ressources génétiques oléicoles de Tunisie].IRESA.

Wahbi, S., Wakrim, R., Aganchich, B., Tahi, H., Serraj, R., 2005. Effects of partial root-zone drying (PRD) on adult olive tree (Olea europaea) in field conditions under arid climate. I. Physiological and agronomical responses. *Agriculture, Ecosystems and Environment*, 106, 289-301.

Wullschleger, S. D., Meinzer, F. C., Vertessy, R. A., 1998. A review of whole-plant water use studies in trees. *Tree Physiology,* 18, 499-512.

INDEX

A

B

C

D

H

I

J

K

L

M

N

O

P

R

S

T

U

V

W

X

Y